Vacuum Bazookas, Electric Rainbow Jelly,

and 27 Other Saturday Science Projects

Neil A. Downie

Illustrations by Jim Wilkinson

PRINCETON UNIVERSITY PRESS
Princeton and Oxford

Published by Princeton University Press, 41 William Street, Princeton, New Jersey 08540

In the United Kingdom: Princeton University Press, 3 Market Place,
Woodstock, Oxfordshire OX20 1SY

Library of Congress Cataloging-in-Publication Data

Downie, N. A. (Neil A.)
Vacuum bazookas, electric rainbow jelly, and 27 other Saturday science projects / Neil A. Downie.
p. cm.
Includes bibliographical references and index.
ISBN 0-691-00985-6 (alk. paper) — ISBN 0-691-00986-4 (pbk. : alk. paper)
1. Science—Experiments. I. Title.
Q164 .D69 2001
507′.8—dc21
2001036258

British Library Cataloging-in-Publication Data is available

This book has been composed in Sabon, Lucida, and Centaur by Princeton Editorial Associates, Inc., Scottsdale, Arizona

Printed on acid-free paper. ∞

www.pup.princeton.edu

Printed in the United States of America

10 9 8 7 6 5 4 3 2
10 9 8 7 6

ISBN-13: 978-0-691-00986-5

ISBN-10: 0-691-00986-4

Contents in Brief

TRANSMISSIONS WITH OMISSIONS

CLOCKS WITHOUT CUCKOOS OR QUARTZ

CURIOUS CONVEYANCES

ANTEDILUVIAN ELECTRONICS

ELECTRIC WATER

INFERNAL INVENTIONS

Contents in Detail

KINETIC CURIOSITIES

STRONG STRING THINGS

STRONG NOTHING

SOUNDS PECULIAR

TRANSMISSIONS WITH OMISSIONS

CLOCKS WITHOUT CUCKOOS OR QUARTZ

CURIOUS CONVEYANCES

ANTEDILUVIAN ELECTRONICS

ELECTRIC WATER

INFERNAL INVENTIONS

Project Ratings

		Difficulty	*Cost*
KINETIC CURIOSITIES			
1	*Hovering Rings*	☆☆☆	☆☆
2	*Dynabrolly*	☆☆	☆☆☆
3	*Gravity Reversal*	☆☆	☆
4	*Maypole Drill*	☆☆	☆
5	*Rotarope*	☆☆	☆☆
STRONG STRING THINGS			
6	*String Nutcracker*	☆☆	☆
7	*Twisted Sinews*	☆☆	☆
STRONG NOTHING			
8	*Vacuum Muscles*	☆☆	☆
9	*Vacuum Bazooka*	☆	☆
SOUNDS PECULIAR			
10	*String Radio*	☆	☆
11	*Mole Radio*	☆	☆☆
12	*Bat Doppler*	☆☆☆	☆☆

		Difficulty	Cost

TRANSMISSIONS WITH OMISSIONS

13	*Toothless Gearwheels*	☆☆	☆☆
14	*Flying Pulleys*	☆	☆

CLOCKS WITHOUT CUCKOOS OR QUARTZ

15	*The Crank and the Pendulum*	☆☆	☆
16	*A Symphony of Siphons*	☆☆	☆☆
17	*Bernoulli's Clock*	☆☆☆	☆☆☆

CURIOUS CONVEYANCES

18	*Dougall or Vibrocraft*	☆	☆
19	*Follow That Sunbeam*	☆☆	☆☆
20	*Duohelicon*	☆☆☆☆	☆☆☆
21	*Fishy Boat*	☆☆	☆☆
22	*Rotarudder*	☆☆	☆☆
23	*Cable Yacht*	☆☆	☆

ANTEDILUVIAN ELECTRONICS

24	*Beard Amplifier*	☆☆	☆☆
25	*Tornado Transistor*	☆☆☆	☆☆☆

ELECTRIC WATER

26	*Meltdown Alarm*	☆	☆☆
27	*Electric Rainbow Jelly*	☆☆	☆☆☆

INFERNAL INVENTIONS

28	*Binary Match*	☆	☆☆
29	*Ultimate Bunsen Burner*	☆☆☆☆☆	☆☆☆☆☆

Preface

Inside this book you will find demonstrations of a number of unusual phenomena, with explanations and analysis of how and why they work. They are nearly all novel in principle or detail as far as I am aware, and most people will not have seen them before or analyzed how they work. I hope that the novelty of the book will refresh and stimulate the imaginations of readers and maybe provide ammunition for thinking laterally in science and engineering.

The demonstrations in the book largely arose from my search for novel science projects to amuse and instruct the kids of the Saturday Activity Center, a Saturday morning club in my hometown of Guildford, near London. I thought it was worth coming up with something original for the sessions because new experiments were more likely to be remembered by the participants. It is, after all, human nature to remember surprises. The occasions when things went wrong—for example, when the test tube exploded, or when water instead of gas came out of the Bunsen burner—are etched in memory more clearly than a thousand ordinary lessons. For each project, I worked out simple explanations and analyses and devised a way of carrying out the demonstration in a simple way. Over time, I realized that I had some unique material that would be of interest to a far wider audience, and I began writing this book.

Those interested in just trying the experiments can do so. I have, however, expanded the mathematical analysis and explanations that go with each phenomenon, to make the book suitable for a more sophisticated audience. The level of mathematics in the book varies from elementary algebra and geometry to simple calculus. The theoretical physics used centers on mechanics and waves but has to touch on a variety of other more complex subjects, from hydrodynamics to electricity and even chemistry. I hope that the university graduate, even the science specialist, will find the discussion stimulating. Despite this, I hope that the book has retained its roots (in demonstrations for kids), and that intelligent young teenagers will still understand most of the main text of the book.

The demonstrations have been improved in many ways, again making them more suitable for a more sophisticated audience, but without, I hope, making them less intriguing to kids. I have done a bit of foolproofing so that the demonstrations depend less on my own magic touch, and I have made some changes simply to improve the analysis discussion. The activities are now nearly all practical demonstrations you can try yourself without expensive apparatus, although they do vary in difficulty.

I have written the book in an informal style, and I have tried to arrange the text in a way that the reader will find easy to follow. (As Dr. Johnson said: "A man ought to read just as inclination leads him; for what he reads as a task will do him little good.")* I have grouped the demonstrations so that their common phenomena and analysis can be compared and contrasted. Each section of the book has a more-or-less similar order of presentation, which should also make for easier navigation. Where necessary, though, I have varied the format a little to suit the subject under discussion. I have set off the more detailed of my theoretical analysis and explanations clearly marked "The Math and the Science," while ensuring that the remaining text stands on its own. The reader can thus skip the more detailed material if desired, at least at a first reading.

I am fascinated with the "forgotten" science of the past. Some of these experiments, although they came to dead ends in the past, will lead to useful technology in the future, as social conditions change and complementary technologies improve.† The introduction and background of each section include some notes and anecdotes on some of the cul-de-sacs of history that relate to the demonstrations.

New phenomena need new analytical understanding, and this is, I hope, one of the values of this book. It will provide practice in applying mathematical physics analysis principles, practice that will seem more relevant and less abstract because it is applied to real demonstrations and experiments. Also, the novelty of the experiments means that there are no right answers written down anywhere in standard books. I hope readers will therefore feel a little closer to

*James Boswell, *Life of Samuel Johnson* (London: Croker, 1837). Dr. Johnson (1709–1784) wrote the first widely used dictionary of the English language. What is more, he did so single-handedly! His biographer, Boswell, remarks, however, that Dr. Johnson was so poorly rewarded for this prodigious labor that, having completed the dictionary, he was forced to continue writing to earn his daily bread. The publishers' meanness turned out to be a blessing for posterity, because Johnson went on to write dozens more works on a vast array of subjects, from the determining of longitude by magnetic variation to a fairy tale called "The Fountains."

†Many years ago the socialist and scientist J. D. Bernal came up with the concept of "unblocking technologies," new developments that could be combined with devices and other research that had failed historically and might unblock the way toward producing success from failure.

the leading edge of science and will be enthusiastic enough to carry the analyses in this book further on their own.

For nearly all the experiments, the analysis and the simple mathematical models I have given could be extended. First, I have used approximations and simplifying assumptions that the reader might want to modify or improve. The removal of some of these simplifications may well reveal more subtle points in the analysis. Second, the theory of any real phenomenon is always, at the deepest level, more complex than can be fully analyzed with simple algebra. Even where an attempt has been made at an exhaustive analysis, there is usually some of the field left unplowed. (For example, the most ingenious account of the theory of Rotaropes so far published does not take into account air drag or rope twist.) I give only the more superficial levels of analysis for each project and offer some hints as to directions for deeper analysis that might be interesting.

The analysis of nearly all the projects here could be enhanced by computer simulation. Adding numerical simulation to the analyst's armory often allows a more sophisticated treatment, though I believe that some apparently simple phenomena, such as bubbling flow in siphons, would pose a considerable challenge for all but seasoned experts.

I encourage readers to try out some of the demonstrations for themselves. None of them (the Ultimate Bunsen Burner excepted) is particularly time consuming, and they are all rewardingly fascinating or surprising. I am a firm believer in doing both "the thinking and the doing"—practical demonstration is as fundamental in importance as theory, although I might not go as far as Erasmus Darwin (father of Charles "Evolution" Darwin), who once famously said, "A fool . . . is a man who never tried an experiment in his life."

But I would go along with Samuel Johnson once again: "The philosopher may be delighted with the extent of his views; the artificer with the readiness of his hands; but let the one remember that without mechanical performance, profound speculation is but an idle dream, and the other that without theoretical prediction, dexterity is little more than brute instinct."

For readers trying things out for themselves, I have awarded stars to rate the degree of ease or difficulty that the practical work of each demonstration might entail, along with some notion of cost. In addition to a full set of instructions for a basic version of the demonstration that will definitely work, for each project I have also included various hints and remarks about how you might like to extend it. The further developments suggested obviously have a greater experimental content, and although they will be more difficult, they will likely be especially interesting and rewarding.

Finally, if you have suggestions or comments to make, or photographs or drawings of projects, please write or e-mail me care of the publisher. I would enjoy hearing from you.

Vacuum Bazookas, Electric Rainbow Jelly,

and 27 Other Saturday Science Projects

Kinetic Curiosities

I *Hovering Rings*

One Ring to rule them all, One Ring to find them,
One Ring to bring them all and in the darkness bind
them.

—J.R.R. Tolkien, *The Lord of the Rings*

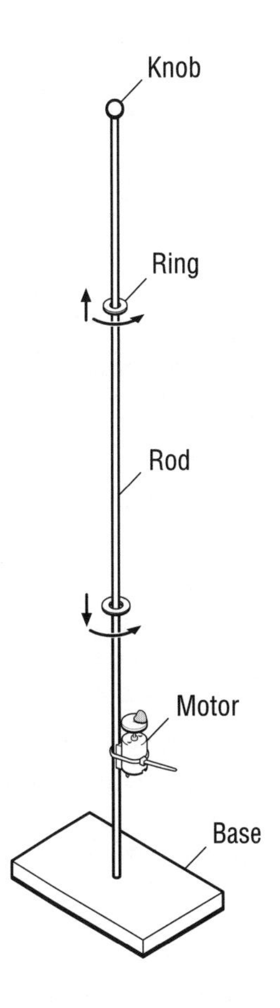

J.R.R. Tolkien's rings have mysterious powers, such as invisibility or the ability to overpower the minds of others. He needed thousands of elves to slave away for years at hot forges underneath mountains to make his rings. Hovering rings are much easier! But the rings here are magical in their own way too: we have rings that go up, rings that go down, rings that go up and down, rings that hover, rings that sink other rings, and rings that bounce off other rings.

A washer-shaped ring of a suitable size, allowed to slide down a suitable rod, will sometimes jam and then spin around the rod rather than simply slide rapidly down. This effect was used in a toy, now long off the market, called Fiddlesticks (described by Jearl Walker in *The Flying Circus of Physics*). Surprisingly, this motion can be enhanced and rings can be made, with the addition of some power to the rod, to go both down and also, in defiance of gravity, up.

The Degree of Difficulty

This is a straightforward project, but you do need a suitable set of parts.

What You Need

- ❑ Round wooden rod (for example, 600–1,500 mm long, 4–6 mm diameter)
- ❑ Wooden base for the rod to fit into
- ❑ Small electric motor (for example, 3 V, 1 A, sold in toy stores)
- ❑ Clamp to attach motor to rod
- ❑ Grommets or glue to prevent vibration damage to wires
- ❑ Batteries, battery box, and wires to power motor
- ❑ Something to control motor speed (optional; see text for details)
- ❑ Wheel to fit the motor shaft
- ❑ Weight to mount on the wheel (such as a lump of modeling clay or a small steel nut)
- ❑ Cellophane tape (Scotch, Sellotape)
- ❑ Small metal and plastic washers, with holes 6–10 mm diameter
- ❑ Small heavy knob (such as an 8 mm diameter dome nut)
- ❑ Nickel chrome resistance wire to control motor speed

What You Do

In the preceding figure, the rod is a round wooden dowel 5 mm diameter and 1,000 mm length. It is supported by a base that is a block of pine measuring 25 × 150 × 250 mm, in the center of which is a hole into which the rod fits snugly. A small heavy knob (such as a dome nut) is attached to the other end of the rod.

On the rod are placed one or two rings, for example, as follows:

Material	*Thickness (mm)*	*Outside Diameter (mm)*	*Inside Diameter (mm)*	*Mass (g)*
Steel	1.4	12	6.5	0.8
Nylon	1	17	8.5	0.3
Nylon	2	22	10	0.6
Acrylic	1.5	26	13	0.7

The vibration source is adjustably attached to the rod by means of a clamp placed about 15–30 percent of the rod's length from its base, with the motor axis parallel to the axis of the rod. When the vibration source is activated, the rod should vibrate with an average amplitude (which varies smoothly along the

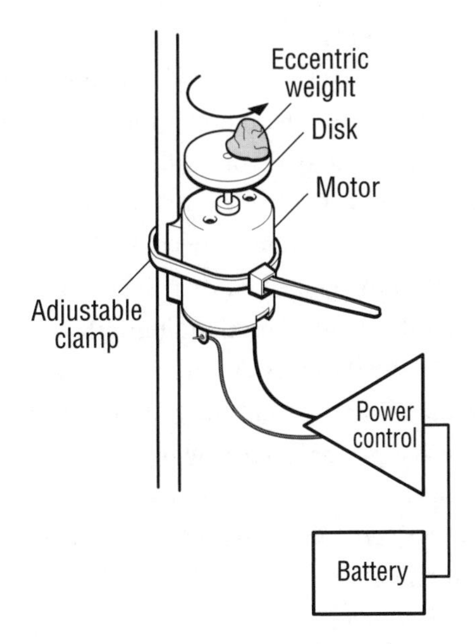

rod, being zero at nodes and maximum at antinodes) of about 1–5 mm from its rest position.

The vibration source is shown more clearly in the figure above. It comprises a disk to which has been attached at a point 10 mm from the axis of the axle a weight of 5 g (this could be as simple as a lump of modeling clay, with some tape to secure it). The disk is driven to rotate around its center by a 1.5–6 V DC motor, the speed of which may be varied by controlling the power supply.

The power supply is conveniently a battery of dry cells whose output is managed by an electronic controller (such as a pulse-width control system). Alternatively, vibration power may be varied by selecting the number of cells to be connected to the motor. I usually used between one and three AA nickel cadmium rechargeable cells on a small electric motor. Another alternative is to use screw terminal blocks to attach a selectable length of 27 AWG, 28 SWG (0.376 mm diameter) nickel chromium wire about 20 cm long. This can be used as a variable resistor of high power-handling capacity.

Now remove the end knob, place one or more washers over the end of the rod, and then replace the knob. Start the motor and see what happens.

The Surprising Parts

The motor will cause the rod to vibrate in such a way that the base and the end knob remain stationary (the nodes), but in between, the rod vibrates with an amplitude maximum in the middle (the antinode). Without a knob on the top, a node tends to occur a quarter or so of the rod length down from the top.

If you tried dropping the washers down the rod before you attached the motor or before you turned it on, you will know that the washers, unless dropped precisely horizontally, tend to catch on the rod and then will tend to spin around and descend the rod rather slowly. As they spin down, they take on a two-lobed appearance, rather like the two fluttering wings of a butterfly in flight, so you might call this kind of motion "butterflying."

With the motor on, after the rings descend the rod, butterflying, they should bounce off the bottom and then ascend the rod until they reach the node, then descend again. What is happening? Has Congress repealed the law of gravity?

Occasionally, a ring will stop going up or down and will stay at a constant height, or it will descend very slowly (taking more than a minute), a mode of motion you might call "hula hooping."

More surprises:

- A ring can bounce off another ring.
- A ring shaped like a bell (or thimble) with a hole in the top will go up, but not down.
- If you interrupt the normal vibration mode by lightly clamping the rod near the top, the rings may not turn back at the node but may instead continue up to the end, where, unless stopped by the knob, they will be flung off at a surprisingly high speed!

Using the Hovering Ring

With only a single ring on the rod, a regular progress both up and down the rod often happens. However, as soon another ring or two are added, the rings seem able to move along the rod in unpredictable ways, still in apparent defiance of gravity, provided only that the motor is running and the rings are spinning. You can alter the pattern of vibration of the rod by touching or pinching it in a number of ways, changing the direction or manner of the rings' travel.

The rod may be removed from its base and held in the hand, with a clamp or suspended by string. Differing means of suspending the rod give rise to different patterns of vibration, in turn causing different behavior patterns in the movement of the rings. Similarly, changes in the power supplied to the vibrating device induce changes in the pattern of vibration and hence also in the behavior of the rings. Other adjustments produce a more highly "chaotic" motion of the rings that is even less predictable.

As well as rings that butterfly, you can add larger but lightweight rings that will themselves only hula hoop, but that are propelled up and down by smaller butterflying rings. For example, try using small steel rings to propel 25 mm or 50 mm (1″ or 2″) paper rings.

The Tricky Parts

The unpowered end of the rod must not be damped too much. It helps to hold it in a clamp such as a car battery charger clip, rather than in your hands, which will absorb too much of the vibration energy. The lead wires to the motor, which are subject to vibration, are supported by extended rubber support grommets (or just a coating of glue) near where they are soldered on, so that they do not suffer fatigue fracture prematurely.

If you let small children play with the device (and in my experience, this gadget is a magnet for small children), try to ensure that the end of the rod doesn't poke one of them in the eye—leave the knob on the end.

THE SCIENCE AND THE MATH

The vibration caused by the motor sets up "standing waves" in the rod. Standing waves are caused when two continuous sets of waves of the same frequency going in opposite directions are set up. The two waves first overlap precisely and add up to a wave of twice normal amplitude, then, as one travels to the left and the other to the right, the amplitude simply shrinks until it is zero: the two waves have canceled each other. So why doesn't the story stop there? Have the two waves annihilated each other? Not quite. Although the material that the wave is traveling through, be it water, string, or air, is momentarily waveless, there is still energy present—that material still possesses kinetic energy, as it is still moving. It is moving just fast enough, in fact, to overshoot the zero amplitude state and form another wave of twice normal amplitude. The situation is precisely described by the mathematics of the "superposition" of waves:

$$Y = \sin(x + Vt)$$

(left-going waves, traveling with speed V)

$$Y' = \sin(x - Vt)$$

(right-going waves, traveling with speed V)

$$Y + Y' = \sin(x + Vt) + \sin(x - Vt) = 2 \sin x \cdot \cos Vt$$

(waves staying in one place but changing size).

Moving waves simply add up, and in this case, they add up to a stationary wave.

Standing waves can be set up by having two sources of vibration pointing at each other, for example, or by having a reflector facing a source. It is the latter case we have here, with the source of vibration the motor and the reflector the end of the rod.

Unlike the normal physical-laboratory standing waves, the waves set up here are not simple transverse vibrations but circular motions, "circularly polarized" waves. These can be considered as two transverse vibrations at 90 degrees to each other in space (and phased apart by 90 degrees in time, of course). Essentially, the entire rod is set into circular vibration, with amplitude and phase varying along the rod, with the exception of the node, which is of course stationary. The node typically is at about 20–30 percent of the length from the top end of the rod (with longer rods, more than one node can be formed).

If you have access to a small disco strobe lamp or, better still, a stroboscope, you can use it to figure out what it going on by inspecting the activity in slow motion. (By the way, you can make a stroboscopic viewer simply by looking through slots in a rotating wheel: a plywood wheel about 8″ in diameter with sixteen slots of 3 mm, mounted on a bearing on a handle, should do.) Another alternative is to silhouette the Hovering Ring against the blank white screen of a computer monitor in a darkened room.

When viewed in slow motion, rings can be seen rolling around the rod on their inner surface, with some slipping and skipping also occurring. The orbiting motion of the rings means that they cannot simply slide down the rod—they can only "roll" around it. Furthermore, the rings are accelerated until they are rolling around the rod at the same speed as the motor is vibrating it. Curiously, they do not normally roll with their plane horizontal, but rather rotate at an angle. When rolling at an angle to the rod's axis, the rings move along that axis. With the correct conditions on the rod, the rings are able to "roll" around the rod in an upward direction, as well as downward.

Unless the power applied by the vibration source is large, the frequency of vibration needs to be near the frequency of a resonant mode of transverse vibration of the rod so that a small amount of applied power results in a relatively large amplitude of vibration. In practice, with a small electric motor giving power, the rod normally vibrates near a resonant frequency. With a reasonable amount of current to the motor, the amplitude of vibration of the rod will be a sizeable fraction of its diameter.

When viewed in slow motion, the pattern of vibration of the rod also becomes apparent: each part of the rod moves in a small ellipse or circle about the rod's axis. The vibration of the rod forms standing-wave patterns, with nodes (positions of minimum average vibration) and antinodes (positions of maximum vibration). The rod shown in the first figure will, in its simplest mode of operation, have two nodes (one adjacent to the base and one at about two-thirds of the rod's length from the base) and one antinode (about one-third of the rod's length from the base). A more flexible and elastic rod stretched between two clamps will have at least two nodes, one at each of the supports. The 1,500 mm rod has a resonant frequency with three nodes at about 30 Hz. A similar rod cut to 850 mm in length also has a resonant frequency of 30 Hz, but with two nodes.

Rods with all fundamental resonant frequencies will also generally have a series of higher resonant frequencies with three, four, five, or more nodes.

The frequency of vibration required to achieve a particular number of nodes increases with the stiffness of the rod (roughly as the square root of the Young's modulus and increasing with the effective diameter of the rod) and decreases as the square root of the density of the rod. There are many good books on the physics of waves, but not all cover waves on rods (see William C. Elmore and Mark A. Heald, *Physics of Waves,* page 113, for a fairly simple analysis of what is actually complex). The selection of frequency is also affected by other factors, such as the end effects due to the supports, the anisotropic nature of the materials (a piece of wood has completely different properties along and across the grain), and any added masses (such as the source of vibration and any knobs fitted to the ends).

The frequencies with smaller numbers of nodes are those of most interest.

The rings will often stay stationary, spinning around the rod in the hula-hoop mode. The analysis of this mode is simple: basically, unless the ring spins sufficiently fast that it can grip the rod, it will slide down the rod. Above a certain minimum spinning speed, however, the hula-hoop speed, the ring will spin in a stable manner around the rod, descending only very slowly over a minute or two or longer.

The minimum hula-hoop angular speed Ω_{min} is given by

$$M\mu \, \Omega_{min}{}^2 \, R = Mg$$

$$\text{that is, } \Omega_{min} = \sqrt{g/(R\mu)}$$

where M is the mass of the ring, μ the coefficient of friction between rod and ring, R the radius of rotation of the ring, and g the acceleration of gravity. If the rod has a small amplitude of vibration, then R can be taken to be simply (inner radius of the ring minus radius of rod, or $R_i - R_r$). However, to this radius must be added half the amplitude of the rod vibration, or a little less, depending on the degree to which the rod is vibrating in a noncircular, that is, roughly elliptical, pattern.

The butterfly motion of the rings on the moving rod is a curious one. The angle at which a ring moves up and down the rod could be dubbed the butterfly angle, B_{max}. It is given by

$$\cos B_{max} = R_i/R_r,$$

which is, of course, the steepest angle at which it can rotate. It would seem that it could rotate at any angle smaller than B_{max}, but in the rings I have studied, it does not do this. It rotates instead at an angle perhaps $B_{max}/2$. Perhaps this has something to do with the fact that the ring is spinning as well as rolling, so its actual spinning speed is not the same as the rotation speed of the envelope of its motion.

When butterflying, the ring is actually spiraling around the rod, rolling its inner surface around the rod. If we assume that the rod is vibrating rotationally at f Hz, then the transit speed V_t of the ring up and down the rod should be given by

$$V_t = 2R_i f \cos B.$$

In a typical set-up, f might be 30 Hz (1,800 rpm), giving, with a 4 mm diameter rod and a ring with an 8 mm inner diameter, a ring transit speed of 200 mm/second.

So rings with a smaller inside diameter should go up and down the rod more slowly, other things being equal, and rings with larger inside diameters should go faster—unless they are too big to go at all.

Hovering Ring devices of various different characteristics may be made to hover or butterfly using these principles.

Rings of a wide variety of specifications may be used on the rod. For example:

Material	*Shape*	*Outside Thickness*	*Inside Diameter** (mm)	*Diameter* (mm)	*Mass* (g)
Nylon	Annulus	2	24	13	0.7
Rubber	Annulus	1.2	27.5	10	0.6
Steel	Annulus	0.8	18	8.5	1.3
Nylon	Annulus	1	61	13	3.1
Polypropylene	Annulus	1	40	21	0.8
Acrylic	Ellipse	2	1/21	13	0.8
Steel	Hexagon	8	14	8	6.6
Nylon	Cuboid	1.5	35/21	13	0.6
Nylon	Square	1.5	22	13	0.4
Nylon	Triangle	1	43	13	1.5
Polypropylene	Hat shape	1.5	30	13	0.8
Aluminum	Ring with lip	0.5	30	13	1.5
Polypropylene	Ring	4	31	12	0.8

*Effective outside diameter (where two dimensions are given, these are the maximum and minimum chords).

The rings may be made of any convenient material, including metal, plastic, and cardboard, and are most visible in bright colors. They may be of a wide variety of shapes, as long as each is pierced (roughly in the middle) by a hole larger than the vibrating rod by from 20 to 200 percent. They may be asymmetrical with respect to the principal plane of the ring. Hollow rings are a good idea, being lighter in weight but visually large. The overall dimensions of each ring may be from marginally greater than the cylindrical hole, up to twenty or thirty times the diameter of the rod.

Although all these rings work individually and many will function well in multiples or in combination with others, certain combinations and multiples will not work. For example, with too many rings on the rod, the vibration power available may be insufficient to operate all the rings. On the rod described here with the motor and batteries specified, only about a half dozen rings of lighter weight can be made to "hover" simultaneously.

Rings that are asymmetrical around the hole with respect to the plane of the ring can also be used, for example, those with an upward-pointing lip or downward-pointing rim (not necessarily a complete rim), as illustrated. The figure shows a cross section of two such rings through a plane passing through the rings' diameter and rotational axis of symmetry. Because the rolling surface is above the center of gravity, these behave differently from symmetric rings. The upward-pointing lip would appear to make travel downward easier than travel upward. However, surprisingly, these rings ascend more easily than they descend, and ascend more easily than do

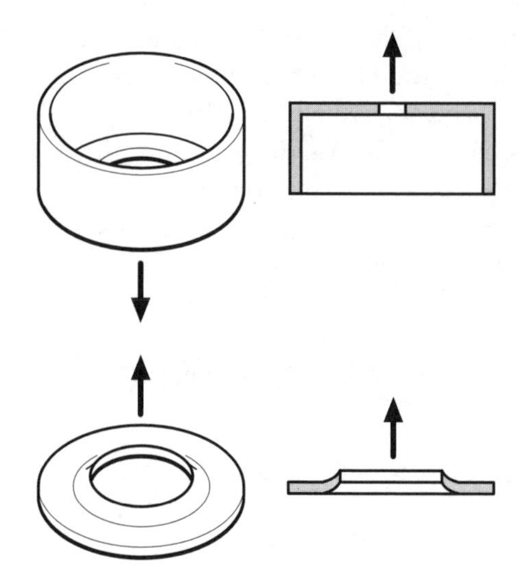

symmetric rings. They typically climb upward more easily than downward (or if turned upside down, descend much more easily than they ascend).

Fast-moving light rings can be made to bounce between slow-moving heavier rings presumably because the impact flips the light ring from downward helical rolling to upward helical rolling. The fact that these light rings seem to go up and go down at about the same speed would lend credence to this hypothesis.

Very large, thin rings (for example, 0.8 mm thick, 50 mm diameter) may have unstable spinning modes when they oscillate to and fro in their rotational axis, giving the appearance of a butterfly flapping its wings.

And Finally, for Advanced Users: 57 Varieties of Hovering Rings

Vibration Sources

Although the simple motor with an eccentric weight is probably the simplest means of providing power, other sources are of course possible, including moving-coil or moving-iron magnetic mechanisms and hand-cranked mechanisms.

An alternative vibration source employs two eccentric (off-center) weights. A 1.5–6 V DC motor with two projecting shaft ends is equipped with two disks with eccentric weights. Each disk carries an eccentric weight. These weights are equidistant from but on opposite sides of the axis of the disks.

Mounting the motor on a moveable clamp allows the vibration source to be moved along the rod. Changing the clamp's position alters the pattern of vibration of the rod, producing a variety of operating effects.

The power required of the vibration source is a function of how well the frequency is tuned to the resonant frequency, the diameter and length of the rod, and the total weight of rings to be used at any one time. The output of an electric motor used in a vibration source may be controlled simply by changing the number of cells in a battery. Alternatively, power can be controlled by rheostat (variable resistor), by electronic pulse-width control, or by series diodes or transistors.

The power of the motor and the mass of the eccentric weight may be chosen so that the device is responsive to the vibrations of the rod, giving the vibrating source an element of self-tuning to one or more of the rod's natural frequencies. A directly coupled motor of low power with a relatively large eccentric weight will exhibit this self-tuning effect. In the absence of this self-tuning effect, the rotation speed of the motor must be tuned to the system. A computer-based system that controls both amplitude and frequency might be an interesting advanced development.

A second source of vibration may be fitted to the rod, thus increasing the aggregate power of vibration and changing its distribution. This is particularly desirable when the rods are in complex shapes—for instance, in an arc—and where the end of a rod may otherwise be too far from the single source of vibration. The two sources of vibration need not operate at the same frequency and power; where their frequencies differ, the amplitude of vibration of the rod is modulated by the "beat" or difference frequency, thus providing another modus operandi.

Rods

More than one similar rod can be activated by a single vibration source: with rings going up and down on two or more rods, the effect is quite interesting. (If the rings are heavy enough to affect the vibration frequency slightly, then the rings on the different rods will interact.)

Rods of different diameter and surface finish offer different sets of behaviors. I recommend rods fitted with an end knob, to avoid eye injuries. Rods can be vertical or inclined. Vertically mounted straight rods provide the most counterintuitive mode of operation. However, rods may be oriented at different angles and may be bent in an arc, although these may require a lot of adjustment to make them work.

The rod may be made of any convenient material, including wood, metal, plastic, and glass fiber. It should be long, thin, with the ratio of its length to diameter being at least 100 (200 is better), and at most about 1,000. The cross section along its length may be constant or tapered. Ideally the rod should be flexible (so that it does not break if bent in a curve whose radius is of a similar order of magnitude to its length).

The pattern of vibration of a rod may be changed by clamping it at any point along its length, manually or with a mechanical clamp. Only small clamping forces are generally needed to change modes of vibration. Any base or device used to hold or clamp the rod or to bend the rod into an arc should not absorb vibration. Thus a metal clamp works better than a thick felt pad.

Methods of Use

There are a number of ways to operate the device, some of which need a lot of skill. In operation, a ring may initially be confined to the region between the drive motor and the first node on the rod, perhaps the first 40 percent of the rod. There are various ways to modify the vibration of a particular rod and vibration source so that rings travel the full length of the rod. As well as temporary clamping at different points on the rod, you can change the frequency or power of the vibration source. For rings with more than one mode of rotation (for example, large rings), the mode of rotation of a ring may be similarly affected by modifying the vibration of the rod, or alternatively by touching the ring momentarily in the correct manner.

The speed at which different rings (with different diameter holes) go up and down the rod is clearly something that should be investigated scientifically. What difference does the weight of the ring make? Naturally, if the rod is very smooth (or oiled!) the rings won't hover. But avoiding this extreme, does the smoothness or material of the rod matter?

The knob may be replaced by a dangling bell, which sounds when struck by an ascending ring. A competitive game may be played with two hovering rings each, with bells arranged so that the movement of rings on one rod can clash over rings on the other rods. With each rod controlled by one player, players try to ring the bell on their own rod while preventing the others from reaching the top.

There are all sorts of stunts you can try with the Hovering Ring. With more rods, you can have races to get the rings over the top. You can bend a long rod around in an arc and, with luck and skill, get the rings to climb up the arc, go down the other side, and sometimes even return. The possibilities are limitless.

The Hovering Ring has been patented. Many people do not understand patents and think they exist just to keep things from being made cheaply. But in fact a patent is there to restrain widespread imitation in the early years of the development of an idea. This encourages investment by manufacturers in research and development, in tooling up, and in promotional sales.

Without a patent, many potentially useful developments end up as "orphan" inventions. For example, suppose Dr. Dogood develops a new cure called Perfecton for brain cancer. He announces it to the world and says that he is not patenting it. Glaxo-Wellcome or Pfizer will say, when approached: "If we invest in a $500 million testing program to prove that Perfecton works, other manufacturers will make it and we will realize no profits to pay back our investment. So we won't invest in the testing program." So Dr. Dogood's invention becomes an orphan and is never developed.

REFERENCES

Downie, Neil A. "Mechanical Toy." International Patent # WO00/35550.

Elmore, William C., and Mark A. Heald. *Physics of Waves.* New York: McGraw-Hill, 1969.

Walker, Jearl. *The Flying Circus of Physics.* New York: Wiley, 1975.

2 *Dynabrolly*

Parasol, a parasol, a most ingenious Parasol, we've quips and quibbles all in flocks, but never such a parasol.

—With apologies to William S. Gilbert and Sir Arthur Sullivan, *The Pirates of Penzance*

Have you ever wrestled with an umbrella in a high wind and been left with broken spokes and a handful of pieces of twisted wire, still attached to a perfectly good piece of cloth and a perfectly good handle? It seems such a shame to throw away the umbrella when only a quarter of it is broken. Surely in the long history of umbrella inventions there is something to deal with this? Well, look into the archives of the world's patent offices (increasingly available on the Internet) and you will find umbrellas galore—umbrellas that fire bullets, umbrellas that function as radio aerials, inflatable umbrellas, the list is almost endless. (Rodney Dale and Joan Gray's *Edwardian Inventions* gives a few of these.)

Perhaps the Dynabrolly is the solution. It has absolutely no spokes, and "what ain't there can't get broke." Whether it will keep you dry in the rain is another matter—and how long would the batteries last?

The Degree of Difficulty

This project is for the most part straightforward. (The many strings required in the related Electric Lasso project, which appears in the section for advanced

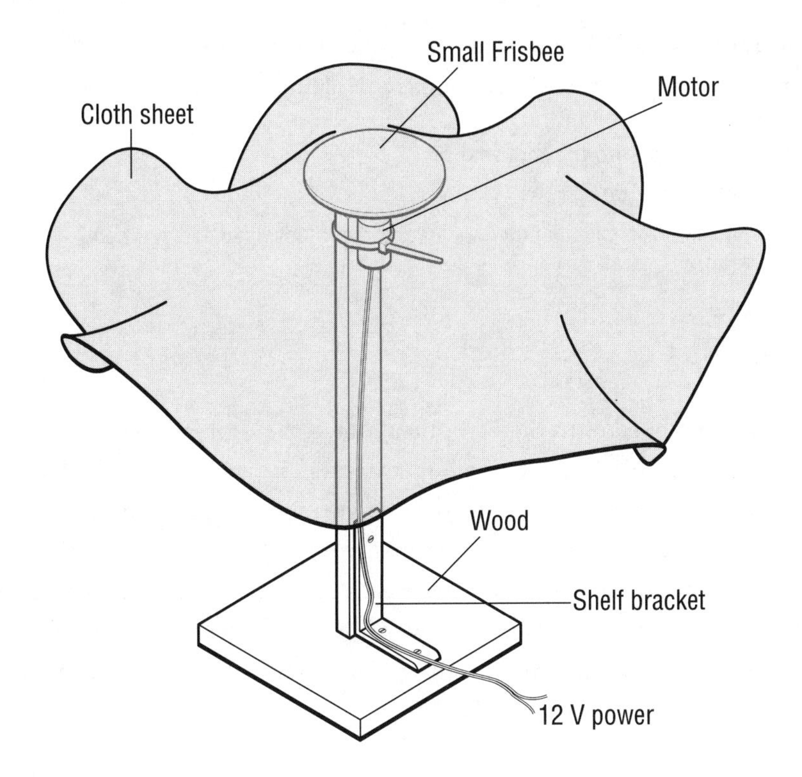

users, make that harder work.) I use a shelf bracket to avoid any problems related to getting the pole upright on the base. You can use two brackets at right angles to make it more rigid—or more simply, you can avoid the problem by sharpening the end of the pole, going outdoors, and pushing it into the ground.

Those with access to antique audio equipment can make a particularly simple Dynabrolly. Simply fit a circle of cloth to the top of a 20–30 cm wooden post placed over the central peg of the turntable of a record player. If the turntable is set to 78 rpm, and the cloth is larger than about 17 cm radius, then Dynabrollying should ensue.

*I used an automobile battery charger and a 12 V motor. About 1 revolution per second will work, but it is best to be able to vary the speed both lower and higher than this. Some battery chargers have a 6 V setting for use with motorcycle batteries as well as the 12 V for cars, which would give two speeds.

What You Need

- ❑ Geared electric motor with suitable reduction gear (see text)
- ❑ Source of power to match the motor*
- ❑ Heavy piece of wood for base
- ❑ Shelf bracket (second bracket is optional)
- ❑ Upright stick

- ❑ Circular piece of medium-weight cloth, about 1 m diameter
- ❑ Small Frisbee, plastic saucer, round piece of plywood, or similar round object
- ❑ Wheel to clamp onto the motor shaft
- ❑ Glue, cellophane tape, bolts
- ❑ Sheet of thin, stiff plastic (such as an acetate sheet used for overhead projection)

What You Do

Attach the shelf bracket to the base, then attach the upright stick to it. Mount the motor on the upright piece of wood, attaching it firmly with tape. Use some double-sided foam tape between the stick and the motor. Attach the motor wheel to the Frisbee with the bolts, then center the round piece of cloth on the Frisbee and glue it on. (I used a 90 cm diameter piece of slightly stretchy light fabric used for infants' sleepwear.) I found it was worth wrapping the motor and transmission in a sheath of plastic and taping this firmly on, so that the cloth cannot get tangled up when you start the motor up.

The Tricky Parts

Make sure that if the cloth becomes wrapped around the central stick, it cannot become entangled. You may find that this does tend to happen when you start the motor. You can obviate this by a gentle start-up. I achieved this in my Dynabrolly by blipping the motor on and off quickly (two or three times a second) to start with. Once the cloth is spinning well, apply full power continuously.

The Surprising Parts

You might expect the Dynabrolly cloth to go around and around in a slightly conical symmetrical sheet. It doesn't! Instead, it forms itself into waves, which go around at half the speed (or so) of the cloth rotation. You can see the cloth moving through the waves if you add spots to the cloth with a marker. If you can use a disco strobe light, you can more clearly see the waves developing, as the flashes of the strobe light "freeze" the motion of the waves momentarily. If you can synchronize the timing of the strobe flashes, you will see the Dynabrolly waves even more clearly.

Using Your Dynabrolly

Try different speeds. You will discover that different speeds tend to give different "modes"—different numbers of waves traveling around and around. Simply starting the Dynabrolly gives different modes sometimes—I can get both three- and five-spoked modes in my set-up at 12 V.

If you turn your Dynabrolly so that the axis is horizontal, and the motor is turning fast enough, you will find that it continues to function as a Dynabrolly, but with some changes in behavior. The traveling waves behave rather strangely, the sideways motion distorting them and tending to crowd them into the upper edge. Sometimes they even begin to take on some of the character of breaking waves at the seaside, rather than regular sinusoidal waves. A horizontal Dynabrolly also has the curious property that it will not rotate (at least at the speed I did tests at) in a vertical plane, even allowing for the waves in it. The disk always veers off to one side or the other to form a cone shape.

The Short Explanation

The speed of a wave on a piece of cloth is governed by the tension in the cloth and its mass. The more tension, the faster the wave; the heavier the cloth, the slower the wave. In this case, the acceleration caused by making the cloth go in a circle causes a force that is exactly proportional to its mass, so heavier cloths behave roughly the same as lighter cloths. You can prove this by wetting the cloth thoroughly: despite the cloth's being much heavier, there will be little change in its shape at speed. (But very light cloths are more affected by the air friction.)

This acceleration is proportional to radius, so the speed of the wave on the outside is faster than that on the inside; the dominant waves formed are shaped like wedges of pie and travel around the cloth in a circle. (Other waves transiently travel in other directions but quickly die out.) The dominant waves on the cloth, it turns out, do not die out but have energy fed into them by the combination of gravity and the air friction as the cloth turns. The effect of the air friction and gravity also leads to the characteristic wave form—rounded on the crests, sharp on the troughs.

The Dynabrolly works the way it does because of a combination of wave effects and its rotational dynamics. The details are very complex, and not analyzed in full anywhere I know. (But try *Vibrations and Waves* by H.J.J. Braddick for a simple account of wave motions, and *Classical Mechanics* by T.W.B. Kibble for a sophisticated analysis of rotating body mechanics.)

The reason the cloth falls into the shapes it does as it rotates is that radial wedge-shaped waves form on the surface of the cloth, and these travel around the cloth at a slower speeds than the rotating speed. Waves on water move without moving the water they are made of (breaking waves on beaches are an exception). If you watch a cork floating on a pond with waves in it, the cork bobs up and down as the wave passes, but it is not carried along by the wave. In a similar way, waves on cloth move without moving the cloth bodily. In this case, the waves travel counterclockwise relative to the cloth, but the cloth is itself not stationary but rotating rapidly in a clockwise direction. The overall effect is that the waves appear to travel around along with the cloth clockwise, but at lower speed.

As mentioned, the speed of a wave on a stretched piece of cloth is governed by the tension in the cloth and its mass:

$$\text{Speed of wave} = \sqrt{T/M},$$

where T is the surface tension in the cloth, and M the mass per unit area.

For an evenly stretched membrane, such as the surface of a musician's drum, this equation applies exactly. However, with the Dynabrolly, things are more complex, because unlike the case of a membrane of a drum, the acceleration needed to make the cloth go around in a circle causes a force, which causes the tension, which is proportional to its mass, so heavier cloths behave roughly the same as lighter cloths.

$$\text{Force} = \text{Mass} \times \text{Acceleration}$$

$$\text{Centripetal acceleration} = \omega^2 R$$

$$\text{so Centripetal force} = m\omega^2 R = \text{incremental radial tension force at radius R},$$

where m is the mass of an element of the cloth, and ω the angular velocity (measured in radians per second). However, the total force at any radius must be obtained by integrating with respect to radius from the rim toward the center:

$$\text{Radial tension} = \int (2\pi R m\omega^2 R)\,dR = \tfrac{2}{3}\pi(R^3 - R_o^3)m\omega^2,$$

which gives, dividing by the length of perimeter over which the tension is exerted, a radial surface tension S_r (tension per unit length),

$$S_r = m\omega^2(R^3 - R_o^3)/3R,$$

where R_o is the outer radius. This is the radial surface tension, which will govern waves in the radial direction, but we wish to study waves traveling in the tangential direction.

Now the cloth will also have tangential surface tension force S_t in it. We can get some idea of the tangential tension by considering the tension in a spinning circular string. If you think about the force on a short segment of the string and equate the radial component of the tension in each direction with the radial force, you will have

$$S_t = RS_r;$$

$$\text{so } S_t \sim \tfrac{1}{3}m\omega^2(R^3 - R_o^3).$$

Thus the tangential surface tension is still largest near the center, going to zero at the perimeter, but less steeply than the radial surface tension. The high tension toward the center means that any waves that form are of low amplitude, while the low tension at the rim allows the formation of waves of much larger amplitude. These waves on the rim pull the cloth from inner rings along with them, and the dominant waves formed are shaped like wedges of pie.

As we said, it turns out that the dominant waves on the cloth do not die out (this is a really complicated part) but have energy fed into them by the combination of gravity and the air friction as the cloth turns. It is also complex to predict the wave shapes, but the shape in this case—rounded on the crests, sharp on the troughs—is also the result of the combination of gravity, rotation, and air friction. However, it is interesting to note that some water waves also have this kind of asymmetric wave shape.

For Advanced Users

Try your Dynabrolly out outside: what happens when the wind blows? And when it rains? I noticed raindrops going upward from the edge of the cloth from the crests of the waves formed—a case of sending the rain back where it came from. Also try attaching weights to the edge of the cloth. With small weights there is little effect until a critical value is reached. With two weights, for example, as the weights are increased, there is a sudden change from the mode of rotation seen before to a mode in which the two weights fly out on the cloth, forming two rodlike spokes with trailing loose skirts, with folds in the two skirts following a spiral pattern.

The cloth need not be a flat sheet: try a cone with different angles. Does the surface form a perfect cone (no waves) above a certain velocity? (When I tried it, the cloth formed itself into a waveless volcano shape, nearly flat at the edges and more pointed in the middle—perhaps a result of the stretchiness of the fabric I used.)

And Finally, the Electric Lasso

Using the same powering device as the Dynabrolly, you can make a rotating loop of rope that does not behave in the same way because the effects of air are much reduced. Using another hub similar to that used for the Dynabrolly (a small disk), attach a set of radial strings that go out to a circular loop of reasonable heavy rope. The rope must be a compromise between weight and flexibility. I tried a piece of chain, which has another advantage: you can count links to get a strictly even distribution of the radial strings around the periphery of the chain. I used a steel chain with 144 20 mm links and 36 radial strings, tied onto holes in the small Frisbee positioned with the aid of a protractor conveniently marked

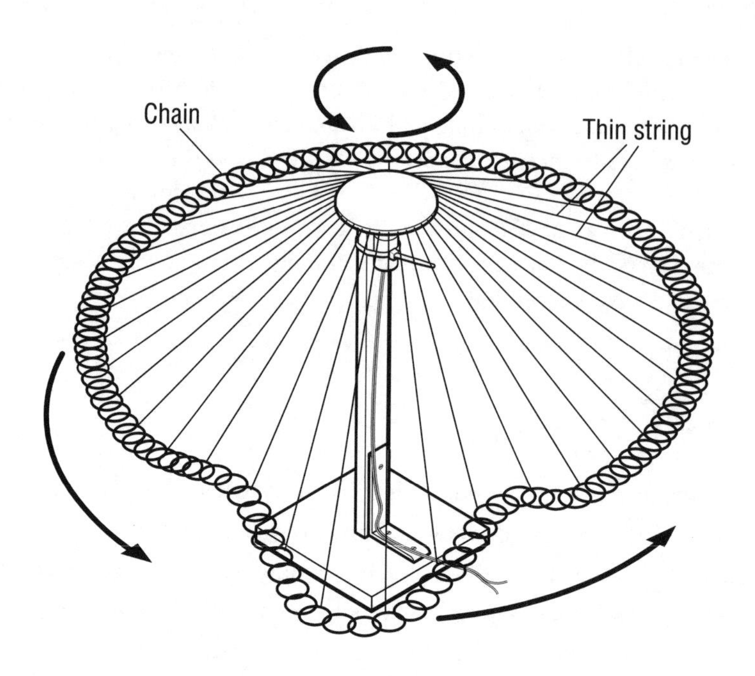

at 10-degree intervals. Waves on the lasso appear to travel at the same speed as they do on the edge of the Dynabrolly. The effect of the air—continuous waves—is removed, however, and waves die out, so that the lasso tends to revolve smoothly in a circle. However, waves can be imposed on the rotating chain by tapping it, and they will continue to rotate for some time around the perimeter.

REFERENCES

Braddick, H.J.J. *Vibrations and Waves.* New York: McGraw-Hill, 1965.

Dale, Rodney, and Joan Gray. *Edwardiaı, Inventions.* London: W. H. Allen, 1979.

Kibble, T.W.B. *Classical Mechanics.* 2d ed. Maidenhead, UK: McGraw-Hill, 1973.

3 *Gravity Reversal*

Inadvertently I made this substance of mine, this Cavorite, in a thin, wide sheet. . . .

Well, so soon as it reached a temperature of 60 Fahr, and the process of its manufacture was complete, the air above it, the portions of roof and ceiling and floor above it ceased to have weight. I suppose you know—everybody knows nowadays—that, as a usual thing, the air has weight, that it presses on everything at the surface of the earth, presses in all directions, with a pressure of fourteen and a half pounds to the square inch? . . .

You see, over our Cavorite this ceased to be the case, the air there ceased to exert any pressure, and the air round it and not over the Cavorite was exerting a pressure of fourteen pounds and a half to the square inch upon this suddenly weightless air. Ah! you begin to see! The air all about the Cavorite crushed in upon the air above it with irresistible force. The air above the Cavorite was forced upward violently, the air that rushed in to replace it immediately lost weight, ceased to exert any pressure, followed suit, blew the ceiling through and the roof off.

—H. G. Wells, *The First Men in the Moon*

Wells imagines a substance with the power to shield the force of gravity, and the disastrous consequences that might easily flow from this. A number of times since, writers have imagined even bolder schemes, where, for example, a government is elected on a platform to repeal the law of gravity. You can just envisage the political commercials: "Are you tired of walking slowly everywhere? Are you tired of carrying heavy shopping bags? Want to build that castle in the air? Vote for me." The problems that would follow if such a thing were possible are of course legion. Careless kangaroos would go into orbit, people on the equator would have to tie themselves onto the Earth, and the candle flames on your birthday cake might snuff themselves out or turn around and burn the frosting! It all goes to emphasize how different a law of nature is from a government law. But is negative gravity possible? Yes, but. . . .

A pendulum normally hangs down from its point of suspension. In this demonstration, with the aid of a ruler and a small electric motor we apparently create a local reversal of gravity—or at least we make a pendulum that will hang upward from its point of suspension. Here, then, at last, is how to repeal the law of gravity!

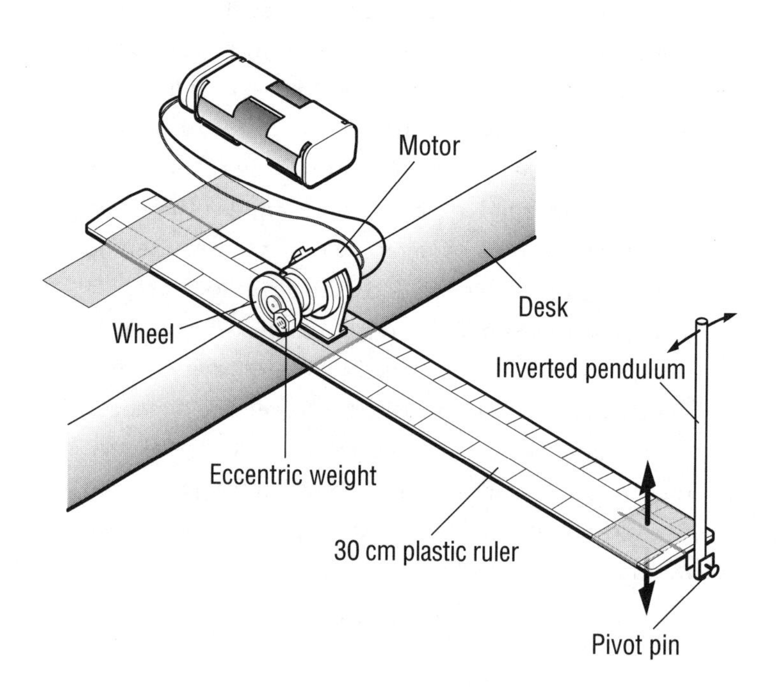

The Degree of Difficulty

The El Dorado of toy companies is a toy that "takes a minute to learn, a lifetime to master." This little gadget is easy to make but very difficult to understand, although I hope that mastering the explanation will not take a lifetime!

What You Need

- ❑ Plastic ruler, 300 mm (1 ft) long
- ❑ Small electric motor (such as 3 V, 1 A)
- ❑ Batteries, battery box, and wires to power motor
- ❑ Wheel to fit the motor shaft
- ❑ A weight to mount on the wheel (lump of modeling clay, small steel nut, etc.)
- ❑ Lightweight pendulum (such as balsa wood or a plastic straw)
- ❑ Pin for pivot of pendulum
- ❑ Cellophane tape
- ❑ Epoxy or superglue and fine sandpaper

What You Do

Insert the pin through one end of the pendulum, then tape the pin firmly to the top of one end of the ruler so that the pendulum can swing freely from it like a propeller on the end of an airplane fuselage. Glue or tape the motor securely to the ruler near the opposite end. The wheel and its weight are jammed on, or glued on with superglue or epoxy (roughen the motor shaft a little with fine sandpaper to give the glue a better grip). Switch the motor on and adjust the distance by which the ruler overhangs the edge of a table or desk, as in the diagram. When the ruler seems to be waggling up and down most furiously, tape the ruler firmly to the table or desk. Then observe what happens when you push the pendulum around from its initial downward pointing.

The Tricky Parts

There really are not many tricky parts. You may need to adjust the weight and radial position of the motor's eccentric weight, and clearly the motor needs enough power to give sufficient vibration amplitude. You can allow for the differences among plastic rulers by adjusting the overhang of the ruler.

The Surprising Parts

When you switch the motor on, the pendulum will hang down in the conventional manner. But if you push it up beyond 45 degrees to the vertically up position, it will leap up to lie at a few degrees from vertical, as if driven by an invisible spring.

The Short Explanation

Take a wooden rod, say half a meter long and 6 mm diameter (but almost any size rod will do), and attach a loose hinge to one end—a flap of tape firmly attached to the end will work. Now place this rod flat on a smooth surface, grasp the hinge so that the rod is free to pivot, and push the end of the rod back and forth along a straight line (the jiggling axis) at an angle to the rod. As you jiggle the rod to and fro, you will find the rod starts to creep away from its original direction toward the jiggling axis. Are invisible imps and leprechauns pushing

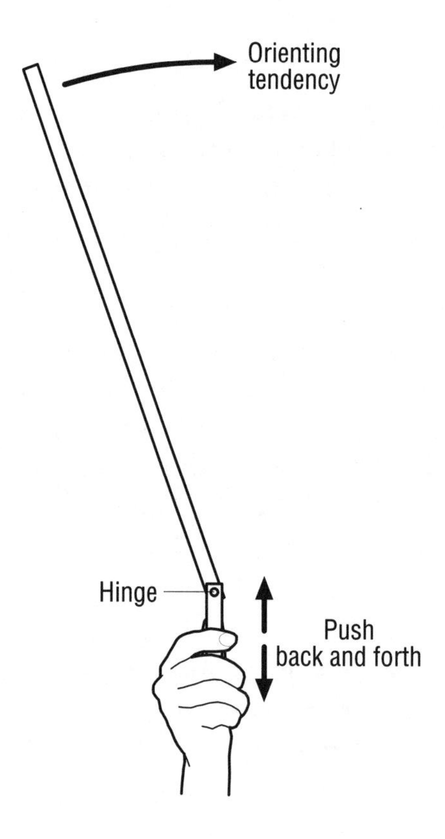

the end of the rod around? No, just simple geometry: with every backward stroke of your hand, the rod's end moves toward the jiggling axis. But every forward stroke of your hand just puts it back again, doesn't it? Not quite: the geometry of the backstroke is slightly different.

I hope you now find it plausible that there is an orienting force toward the jiggling axis. In the inverted pendulum, the situation is even more favorable. Unlike the wooden rod, which stops when you stop the backward stroke, the pendulum can keep on moving toward the jiggling axis after the stroke is finished. The pendulum is therefore even quicker to orient itself. I made a simple computer simulation of the pendulum's movement, using numerical integration, to show the angle of the rod at short intervals of time. This showed the rod moving from side to side like a pendulum, even though it is upside down in the simulation, something that is difficult to see with a highly damped straw or balsa wood pendulum.

THE SCIENCE AND THE MATH

The mathematics that describe this pendulum are relatively difficult, and a bald statement of algebra would give little insight. But one helpful way to think about this gadget is to think about the equivalence of gravity and acceleration. When the pendulum is being accelerated upward by the pin, it behaves as a normal (upside-down) pendulum with a little extra gravity. But upside-down pendulums do not fall over right away: a pendulum takes a finite time to fall, as the following formula shows:

Sideways force $F = Mg \sin \alpha \sim Mg\alpha$
for small angles α,

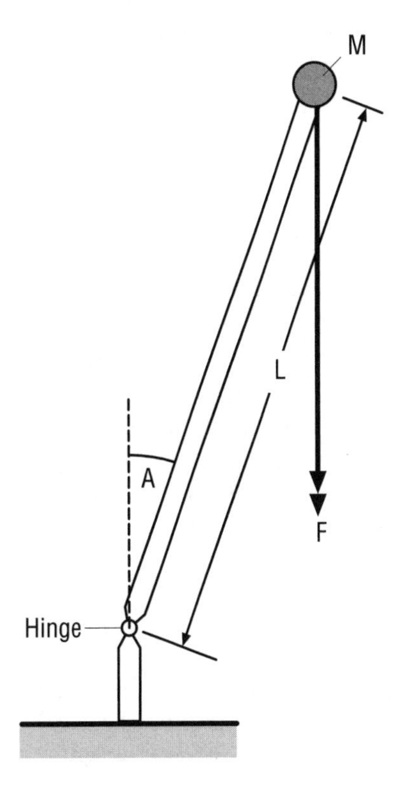

where M is the mass of the end of the pendulum (assuming the rest of the balsa wood weighs nothing), g the acceleration due to gravity, L the length of the pendulum, and α the angle through which it has turned from the vertical. The angle α is given approximately by the displacement distance X from the vertical divided by the pendulum length, that is, $\alpha = X/L$.

Hence acceleration $d^2X/dt^2 = F/M = g\alpha = gX/L$,

leading to a solution

$$X = X_0 \exp(\sqrt{(g/L)}t) \text{ or } \alpha = \alpha_0 \exp(\sqrt{(g/L)}t).$$

X_0 and α_0 are the initial angle and initial position to the vertical (which cannot be zero for any real pendulum).

The formula shows, unsurprisingly, that an inverted pendulum, released from the vertical, tips over at an accelerating rate—but it is a rate with a time constant T which is inversely proportional to the square root of (acceleration of gravity/length of the pendulum). Another key feature to notice is that if the pendulum is started off at a very small angle α_0, it will take longer to fall—but not much longer.

However, when the pendulum pivot is being accelerated downward, it actually has a net negative "gravity" acting on it, momentarily. This negative "gravity" will tend to give the pendulum a restoring force. It is not so easy to show that this restoring force will win over the upsetting force, but it does.

In fact, it turns out that you can think of the pendulum as being subject to a "pseudogravity" field strength G given by

$$G = q(a/2L),$$

where q is the peak acceleration of the pivot, a the amplitude of jiggling, and L the length of the rod. Clearly, the criterion for stability in the inverted position is that the upward pseudogravity has to exceed the local downward gravity and produce a net upward gravity. If you make the rod too long, the acceleration too weak, or the amplitude too small, then the rod will hang only downward.

How real is the pseudogravity demonstrated here? Physicist Ernst Mach declared that gravity and acceleration were indistinguishable, a principle that Einstein used to spectacular effect in his discoveries in theoretical physics. Einstein, in many of his famous gedanken experiments, imagines himself in a spaceship with no portholes and shows that you can't distinguish acceleration from gravity in the theory of relativity. In the same way, if you imagined yourself confined to a spaceship located at the tip of the inverted pendulum, and if you further restrict your senses by making your senses slower than the jiggling speed of the pendulum pivot, then I suspect you will be unable to distinguish the pseudogravity from real gravity in the reversed direction.

It is fairly straightforward to create a simulation of the inverted pendulum using a crude numerical integration. Numerical integration is probably easier than you think. You can use a computer spreadsheet program like 123 or Excel, as I did. I assumed the acceleration of the pivot came in two lumps T_a long in each cycle, a negative and positive, spaced apart in time by T_c. The angular acceleration of the rod depends on the sine of its angle to the vertical; it is zero when the rod is vertical. The angular velocity is obtained by adding kicks, or short impulses, of acceleration (acceleration times T_a) to the previous velocity. The angle itself is obtained by a further process of adding increments (velocity times cycle time T_c) to the previous position.

This simulation will certainly allow you to see the pendulum oscillating to and fro in the inverted position. You may also see with a more sophisticated simulation some of the more complex phenomena, such as the "nodding" oscillations.

And Finally, for Advanced Users

In the version already described, the straw or balsa wood has a lot of air drag, and this means that, like a downward-pointing pendulum made with a lightweight rod, it tends to hang straight after a moment or two—although in this case, of course, it "hangs" straight up!

However, if you use a thin wire rod, and even more if you attach a small weight to the end of the rod, not only will you see negative gravity, you will also see the inverted pendulum swinging to and fro. Measure its period, and you have measured how much net negative gravity you have made, of course!

A more spectacular (but much less safe) way of demonstrating the inverted pendulum—perhaps more suitable for a lecture—is to convert an electric saber saw (jigsaw) to act as the moving support at the bottom. You simply remove the blade and use the mounting screws to attach a long rod with a lug at its base. The saw is powerful and runs on domestic electric power, so watch out for safety issues.

Various researchers—serious people with a zany streak, I suspect—have discovered a number of interesting features about the inverted pendulum's behavior over the years. For example, it will often "nod" as well as oscillate back and forth: the pendulum pauses briefly at different angles either side of vertical (still swinging slightly), generally in a symmetrical pattern relative to vertical. I have seen hints of this behavior in some of the inverted pendulums I have tried with small weights on the end (when the air drag effects are reduced).

The phenomena displayed by variations on the inverted pendulum have also fascinated academics for years. There are learned papers on inverted pendulums of all sorts—those with more degrees of freedom, towers of stacked pendulums made of two or three or more loosely coupled links, pendulums driven at very high frequency, pendulums driven with nonvertical driving, pendulums operated in a vacuum, and so on.

However, despite all the work that has gone into it, it would appear that nobody has thought of a single practical application for the inverted pendulum. The biggest challenge, therefore, bigger even than analyzing how it works, is to find a use for it!

REFERENCES

Acheson, D. J. "Multiple-nodding Oscillations of a Driven Inverted Pendulum." *Proc. Roy. Soc.* (A) (1995): 89–95.

Kalmus, H. P. "The Inverted Pendulum." *Am. J. Phys.* 38 (1990): 874–878.

Yorke, Ellen. "New Analysis of Inverted Pendulums." *Am. J. Phys.* 46 (1994): 285–288.

4 *Maypole Drill*

"And now," as they used to say on *Monty Python's Flying Circus*, "for something completely different." Here is—in contrast to the completely useless previous project—something that is completely practical and actually in active use in less wealthy countries today.

Long before the Black & Decker Corporation provided us all with that ubiquitous household machine, the electric hand drill, quite possibly in the time of the early pharaohs of Egypt, one-hand-operated drilling machines were in use that included no machined bearings, no gears—not even a metal part except the drill bit itself. And they still work very well today.

This is how what I call a Maypole Drill is made, using twisted string around a pole and flywheel.* It can, with a little practice, be as efficient as an electric gadget for drilling small holes. A double version of the drill is almost unique among high-speed rotating flywheels, as it needs no real bearing surfaces. The double version could operate a small grinding wheel or rotary metal wire brush.

*I call it a Maypole Drill because it reminds me of the maypole used in traditional dances still seen occasionally in schools, where the children dance around a pole carrying ribbons trailed from its top, weaving a kind of tartan of ribbon down the pole.

What You Need

- ❑ Round wooden dowel, 300 mm long, 4–5 mm in diameter
- ❑ String
- ❑ Tape
- ❑ Yoke, 100 × 30 mm plywood, about 6 mm thick
- ❑ Nylon tubing, 6 mm size, 30 mm long
- ❑ Twist drill (2–3 mm is a good size to start with)

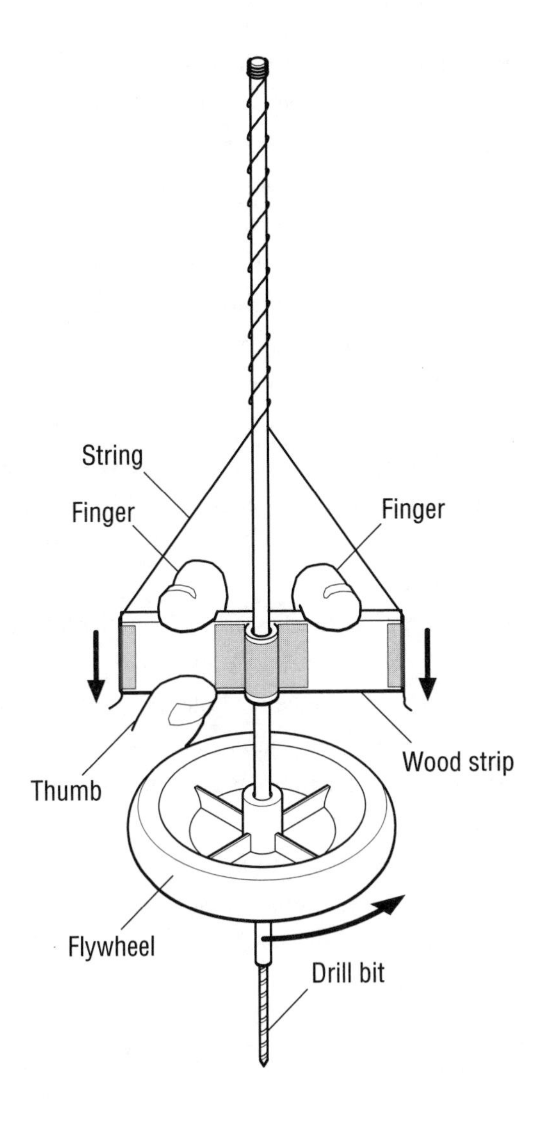

What You Do

Drill a hole in the bottom of the pole and glue the twist drill in the hole. Tape a loop of string around the top of the pole and attach it to the ends of the yoke, so that when the strings are stretched out straight the yoke is still a couple of inches from the end of the twist drill. The piece of nylon tubing is a guide, rather than a bearing, and is fixed in the middle of the yoke, so that the pole can turn freely in it.

Having assembled the drill, position it pointing vertically downward on a piece of wood to be drilled. Hold it in one hand and wind the string up a little with your other hand, and you will find the yoke pulled up. Push down momentarily but hard, and the flywheel will whirl around, rotating the drill and unwinding the string. The flywheel then continues to whirl around until it has wound the string up again with the opposite twist.

Although there is a knack to using the device, after some practice I found that I could easily drill holes from 1 to 3 mm in diameter. The drill offers some genuine advantages over common workshop models. The standard small hand drill is the wheel brace, but with a small wheel brace you need two hands, and it is easy to exert too much sideways force on the drill bit. Unless you are careful to cancel the torque you exert on the wheel with the torque you exert on the handle, it is easy to break the drill bit. Similarly, the standard Black & Decker–style portable domestic electric drill is too large and heavy for the smaller sizes of twist drill, and again the drill bits are easily broken simply because of the unwieldiness of the machine.

On a recent trip to Egypt I noticed many artifacts with holes that could well have been made by drills of this type, and Nick Thorpe and Peter James remark in *Ancient Inventions* on the fine needlelike holes drilled in seven-thousand-year-old obsidian jewelry from Iraq. T.G.H. James's book *An Introduction to Ancient Egypt*, however, insists that the technology used then was the less sophisticated bow drill, and James cites a 1570 B.C. fragmentary tomb painting illustrating one of these. The bow drill seems inferior to me because it needs two hands (three would make it even more manageable) and because the powerful sideways force it exerts endangers the drill bit.

THE SCIENCE AND THE MATH

There are strict limits to the operation of the device:

- If the shaft is too thin, it will whirl around and break as the device is operated (see the comments on whirling failure in the Rotarope project).
- If the shaft is too thick, there will not be enough turns of string on it, and the device will not pick up enough speed.
- If the user presses too hard toward the end of the "spin-up," the drill will dig in and stop.
- If the flywheel is too large, the drill will not reach a reasonable operating speed.
- If the flywheel is too small, the drill will dig in and stop in the wood, not rewinding the string for the next stroke.

The torque given by the string varies as follows:

$$T = 2RF \sin \alpha \cos \alpha,$$

where R is the shaft diameter, F the force applied by the user, and α the angle that the string makes with the shaft. The angle α varies from α_0 (about 45 degrees) to α_{min} (given by $\tan \alpha_{min} = H/L$, where $2H$ is the length of the push bar and L the shaft

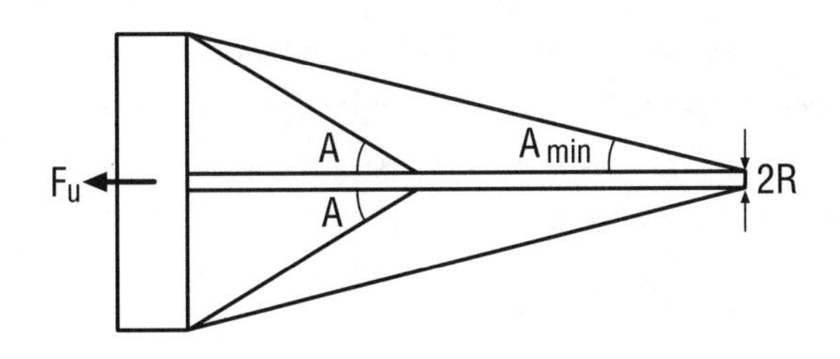

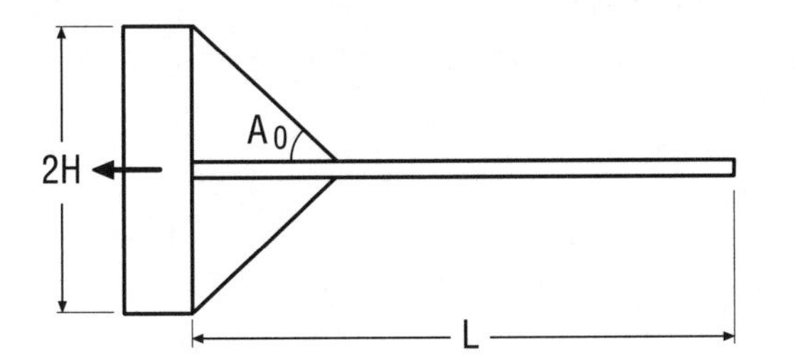

length to the handle as the string unwinds). The torque that can be applied is thus small at the start of the stroke, and this is why it is important to push down hard at first to accelerate the wheel swiftly.

The string itself forms a variable-angle helix, where the helical angle $(90 - \alpha)$ varies smoothly, and where the number N of turns per centimeter varies from a minimum at α_0 to a maximum at α_{min}.

The speed (ω in radians per second) to which the flywheel can be accelerated is probably most easily calculated by an energy balance. Supposing the flywheel to be a flat disk, and the force F exerted for a distance of $L/2$, the energy in must equal energy out to the flywheel:

$$\tfrac{1}{2}FL = \tfrac{1}{2}I\omega^2.$$

The moment of inertia I of a flywheel of mass M with all its mass near radius R is given by $\tfrac{1}{2}MR^2$. So

$$\omega = \sqrt{FL/\tfrac{1}{2}MR^2}\ .$$

Given an F of 50 N (the force equivalent to a 5 kg weight) and an L of 20 cm, R of 5 cm, and M equal to 200 g, we have

$\omega \sim 45$ radians/sec, or 7 revs/sec (420 rpm).

And Finally, for Advanced Users

Now that you have tried the Maypole Drill, why not try the grindstone based on similar principles?

As with the drill, no bearing is required simply to rotate the assembly fast; the double-sided construction avoids the need for one. The pieces of nylon tubing

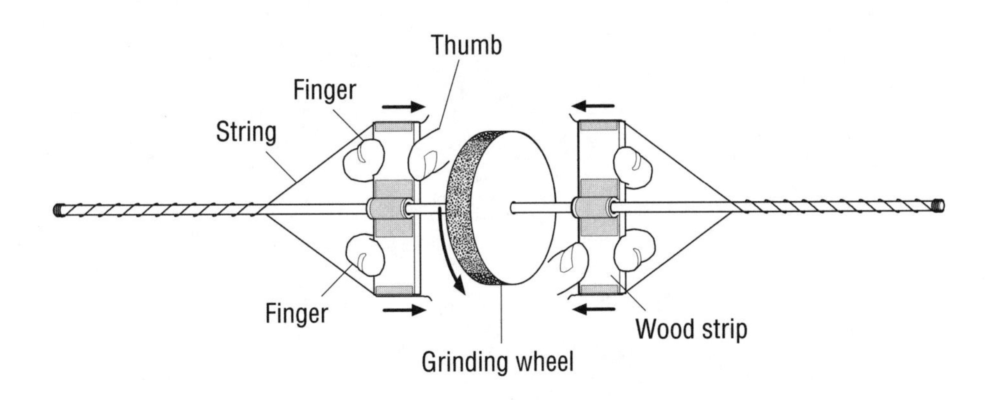

are required as bearings only to take sideways force when the small grindstone is forced against the work piece. Clearly, as with the drill, there is a knack to using the grindstone: you need to synchronize pressure between it and the work piece so that you do not slow it down too much.

REFERENCES

James, T.G.H. *An Introduction to Ancient Egypt.* London: British Museum Publications, 1979.

Thorpe, Nick, and Peter James. *Ancient Inventions.* London: Michael O'Maran Books, 1995.

5 *Rotarope*

The rather formidable theoretical structure that can be based on a study of the motion of such a simple object as a flexible string under tension.

—William Elmore and Mark Heald, *Physics of Waves*

It is a curious thing that any book on waves always seems to start with transverse waves on strings. Are all scientists who are interested in waves amateur violinists, like Einstein?

Curiously, too, the writers of books on waves always claim that a string or rope is a simple object—and then go on to prove just how complex it is! The "elementary" text quoted in the chapter epigraph deploys forty-nine pages of formulas, graphs, and densely packed words—and the analysis would be longer but for the numerous simplifying assumptions made by the authors. The truth is that real things are never simple, and even a piece of string is a real thing and thus not simple.

With Rotarope, we set up a demonstration that illuminates the failure of a rotating shaft, a disastrous occurrence in industrial mechanical power transmission. A significant obstacle in any rotating-shaft power transmission system is the possibility of instability as rotation and vibration modes link undesirably together. Once a shaft is rotating with a slight bend in it, the centrifugal force* on the part of the shaft farthest off center can be enormous, leading to further bending, leading to increased centrifugal force, and so on, leading rapidly to

*Centrifugal force is the imaginary force that is apparently exerted by a body when it is trying to escape from being swung around in a circle. Some may, with good cause, prefer the term "centripetal" force, to describe the force that you need to exert on a body to keep it going around in circles: the centripetal and centrifugal forces have the same value, of course, but point in opposite directions.

catastrophic failure. D.R.H. Jones's *Engineering Materials 3*—like many practical engineering texts—precisely describes the failure of shafts, technically called "whirling" failures. With Rotarope, we replace the slightly flexible steel driveshaft with an even more flexible object—a rope—to see what happens.

The Degree of Difficulty

At the minimum, all you need for this demonstration is a jump rope and some skill in turning it by hand. However, it is difficult to keep turning a jump rope continuously with the same speed and amplitude. I think that it is worth the slight extra trouble to set up the demonstration using electric motors, so that you can "hold" particular patterns of rotation to examine them more carefully than is possible with the transient nature of jump rope patterns.

What You Need

- Wood for mountings
- Two electric motors: these can be the usual sort of small toy or tape cassette motors for a small Rotarope, or larger DC motors for a larger device
- Variable-speed drive for motors
- Batteries and battery box
- Rubber band, string, or thin flexible shaft
- Stroboscopic lamp, stroboscopic viewer (optional)

What You Do

Arrange the motors so that both rotate the same direction, clockwise or counterclockwise. The "rope" can be slightly stretched rubber, string, or, with a large spacing between the motors, a thin shaft.

Switch on the motors. The shaft or rope will immediately deviate from the axial position, forming one or more sausage shapes in the air. To ensure a start at a suitably low speed, adjust the variable-speed drive to give a single sausage shape, a single antinode in the middle—"jump rope" mode. Now speed up the motors and see if a double antinode or "double sausage" mode appears. You may need to help the formation of this mode by pinching the rope in with your fingers.

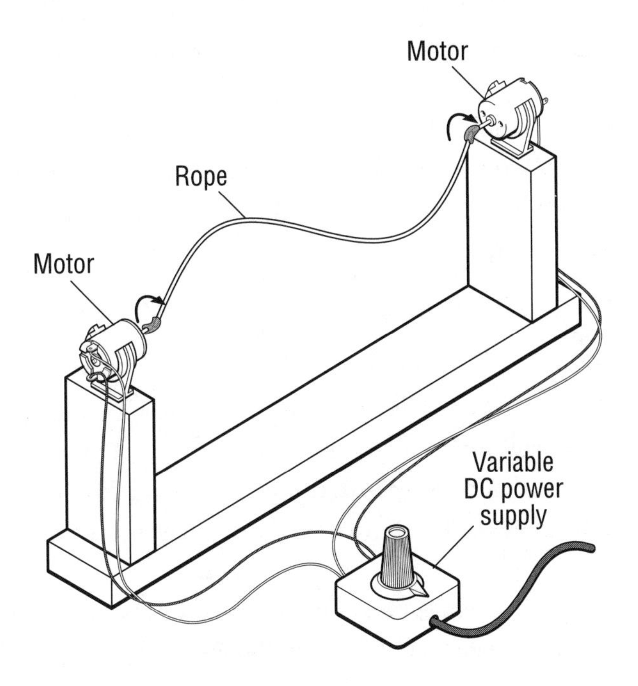

The Tricky Parts

In the standard horizontal layout, because the motors reach the low-frequency "jump rope" mode first, it is difficult to reach the higher modes, until you realize that a little support from your fingers to limit the amplitude of the lower modes will allow the rope to get to the higher modes.

The Surprising Parts

I was surprised when anything other than the jump rope mode appeared. I had expected the rope or shaft to take up a shape somewhat like a distorted version of the catenary curve, the curve of a suspension bridge, and then simply to whirl around in a stable fashion. Robert Banks, in *Towing Icebergs, Falling Dominoes,* describes this mode and calculates its mathematical form, what he calls the "troposkein."

Had I been a child, or more dedicated to fitness training, I would have suspected that these higher modes existed. Anyone who has spent enough time with a jump rope will have realized that it can behave in more curious ways. Rotarope modes can be seen in an even more simple, although transient and unstable, way by simply waggling a jump rope. Certainly one-, two-, and three-"sausage" modes should be visible.

In a related situation, the rotated vertical string described below, T. C. Lipscombe, M. P. Silverman, and Wayne Strange describe in "String Theory" not just the lowest mode, but also the second- and third-order modes of rotation and their mathematical form.

Using the Rotarope

By using the variable-speed drive, you can set the Rotarope to produce other modes of vibration and rotation—mixtures of the first two modes, for example, and higher modes too. At some speeds, the modes will modify continuously from one to another. In some modes the shaft or rope forms into a helical shape while it rotates.

The use of a stroboscope, either a pulsed lamp or a slotted disk, will reveal what is going on in these higher and mixed modes much more clearly.

THE SCIENCE AND THE MATH

The Rotarope exhibits different sorts of vibration modes, depending on the speed of rotation and the length and stiffness (or tension) of the shaft or rope.

The modes depend partly on the speed of transverse waves along the shaft or rope. These transverse waves are generally coupled together to form rotating waves. The addition of a sine wave variation in the vertical plane with a cosine wave in the horizontal plane leads to a circular motion of any particular point on the shaft or rope.

Transverse waves on a stretched string travel at a speed c_s simply given by

$$c_s = \sqrt{T/m}$$

where T is the tension and m the mass per unit length of the string.

Transverse waves on a rod are much more complex, their speed depending on their wavelength and on the cross section of the rods, but on round rods they travel at about C_r, given by

$$C_r \sim (r/\lambda)\sqrt{Nr^2/m}$$

where r is the radius of the rod, N the shear modulus, and λ the wavelength of the wave. In this form, the equation can be compared with that for the speed of transverse waves in a bulk solid:

$$C_r = \sqrt{N/\rho}$$

where ρ is the density of the material. Crudely, the speed of waves on a rod increases when the rod is stiffer, either through a higher modulus of elasticity or through a stiff design of the cross section. So waves go faster on a tube than a rod of the same material and mass per unit length, and faster on aluminium than on plastic.

The Rotarope is not, however, as simple as the regular tight-string vibration mode system—it does not have a well-defined constant tension along its length. For example, a particular average tension becomes established by centrifugal force only as a mode settles into operation. The normal string vibration pattern is modified by the centrifugal force, which applies primarily to those parts of the shaft or rope farther from the axis.

The general analysis of the rope is difficult (see Banks), but a simplified model offers insight. A crude model of the Rotarope appears in the accompanying diagram, where half the rope is considered to be a "weight" and the other half is considered to be the suspension.

The centrifugal force F_c on the weight part is given by

$$F_c = m\omega^2 r/2,$$

where m is the mass of the rope, ω the angular rotation speed, and r the radius through which it swings. The radius r is in turn given by

$$r = (L \sin \alpha)/2,$$

where L is the length of the rope and α the angle that the suspension makes with the axis.

The force F_s on the weight due to the tension T in the suspension is given by

$$F_s = 2T \sin \alpha.$$

Now when the system is rotating at equilibrium, $F_s = F_c$, so

$$2T \sin \alpha = m\omega^2(L \sin \alpha)/4,$$

that is,

$$T = m\omega^2 L/8.$$

This gives rise to the curious result that the tension T in the rope is roughly independent of the angle α, that is, provided the rope is a little slack, the tension when it is rotating will be the same. If the rope is very slack, then the angle of the rope is such that the tension acts more directly, whereas if the rope is less slack, the angle of the rope is such that only a small component of tension acts on the weight, but

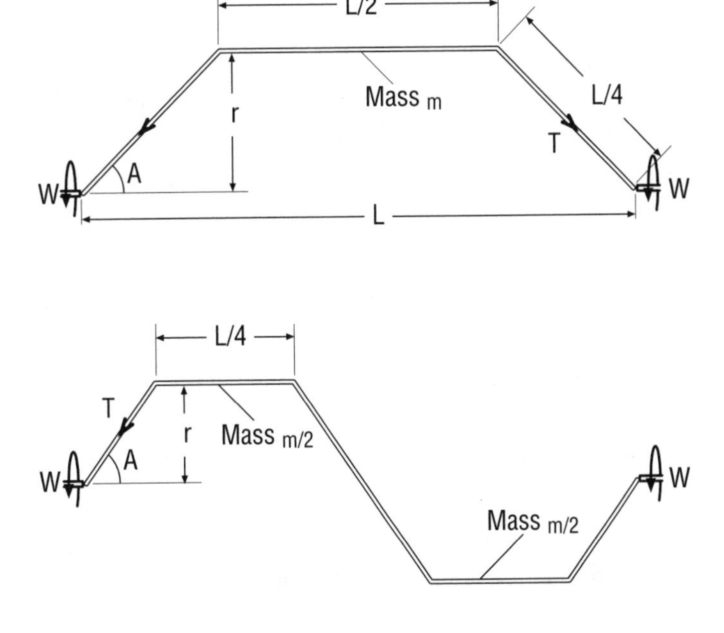

that weight has only a small force on it because the radius of rotation is smaller.

The energy E stored in the motion of the Rotarope is proportional to the square of the angular rotation speed ω and the square of $\sin \alpha$:

$$E = \frac{1}{4} m L^2 \omega^2 \sin^2 \alpha.$$

The different one-sausage, two-sausage, and three-sausage modes of rotation of the rope have an associated energy, due to their different values of ω.

At higher speeds the rope tends to form spiral shapes inside their sausage-shaped envelopes. This is almost certainly an effect due to air drag. In the limit, if you take a piece of light cotton thread and rotate it rapidly, it tends to form a helix with a bulbous envelope shape. With a light thread, effects due to air drag will dominate the behavior.

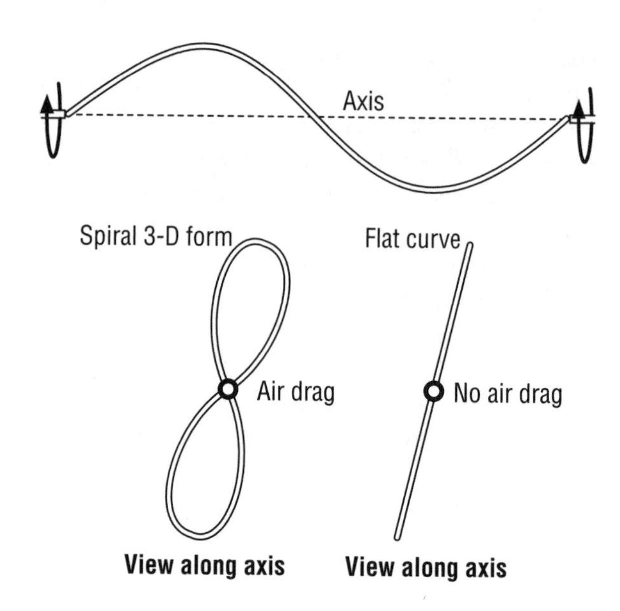

If you have forgotten some of the wave theory you used to know and find the analysis rather daunting, you might like to dip into a book on waves that I found some years ago to be a simpler-than-most introduction to the subject, H.J.J. Braddick's *Vibrations and Waves*.

And Finally, for Advanced Users

Even more than with the Hovering Rings, in this experiment a stroboscope allows you to see more clearly what is going on. A disco strobe light is a very cheap and powerful gadget that is readily available from about $20. Many domestic fluorescent tube lighting luminaires give light in pulses at 120 Hz (100 Hz in Europe), which can also give something of a stroboscopic effect. The effect is sufficiently strong under some conditions that fluorescent lighting of this kind was banned some years ago from workshops using machine tool and power woodworking tools. The rapidly rotating high-speed drills and cutters of these machines can appear stationary in blinking fluorescent light and can tempt incautious workers to reach out and touch them—with disastrous consequences. Some fluorescent lamps show this effect and also show colored bands, which derive from the differing fluorescence lifetimes of the phosphors used in the lamps.

An alternative is to make a stroboscopic viewer by cutting slots in a plywood disk and spinning the disk by hand or via a small, geared electric motor. A "sheet" of light, produced perhaps by a cylindrical lens and laser pointer, or more simply by a slot cut into the mask, illuminated with a small lamp, is another alternative source of illumination that is helpful in viewing the Rotarope. If the sheet of light is positioned to shine along the axis of rotation, then the profile of the rope as it rotates shows up more clearly.

You can try making the shaft or rope longer. You can try adding permanent supports, or temporary supports that you remove once the Rotarope is up to speed: these are the sorts of expedients that engineers in practical applications often have to resort to, in order to avoid serious vibration problems and even failure. What is the largest number of antinodes you can fit between the motors?

With the use of shafts for the transmission of power, the critical frequencies of different modes of vibration (and therefore failure) of rotating shafts have been carefully measured in the past. Engineering handbooks will tell you, for example, that the critical frequencies f_n of the nth mode of ordinary shafts running between universal joints (as used on trucks, for example) are given by the formula

$$f_n = A(n/2L^2)\sqrt{EI/m}\,,$$

where A is a constant of order unity, L the length of the shaft, E the modulus, I the second moment of area of the shaft cross section, and m the mass per unit length. For a cylindrical shaft of radius R,

$$I = \pi R^2/4.$$

These frequencies are often far above normal operation speed, although designs that operate above the first few f_n are possible, provided the machine concerned swiftly accelerates through the resonances.

More Rotaropes: The Vertical Axis Rotarope or Helicoseir

Using only one motor, it is easy to show a related set of behaviors from a vertical suspended rotating rope with a loose end at the bottom. Attach one end of a flexible rope to the shaft of a motor oriented vertically downward, and let the other end hang down free, and then run the motor, beginning with fairly slow rotation speed. What happens is a function of speed. At low speed, the rope will form a simple arc that describes a sausage with a downward-pointing bowl below as it rotates. At higher speeds, the rope will form a more complex shape that will form several sausage shapes in the air ending with the downward-pointing bowl as it rotates. Be sure to employ a relatively "heavy" (that is, high-density, but flexible) rope or chain, to reduce the air drag effects seen with lighter strings. The length of rope or chain needed is subject to experiment, but I used a piece of chain of the kind used for bathtub plugs, which could easily be coiled into three or four turns, giving it the necessary flexibility to show higher modes. I also used a geared-down motor for the best effect: a 10:1 reduction ratio, although direct drive also works.

The Helicoseir has a tension in the rope that increases as you go upward along the rope, thanks to gravity (the analysis of the Rotarope ignores gravity). According to the theory of vibrations on strings just noted, then, the speed of waves will increase as you go up the string, the speed increasing as the square root of tension. Qualitatively this means that you would expect to see more waves crowded into the bottom of the rope, where the same waves are traveling more slowly, and hence there are more of them. Try it and see!

Modes that are very complex are readily possible in the vertical Rotarope—I thought that I could detect as many as six antinodes under the right conditions, but with considerable mode mixing. Modes shift when the axis of the motor is

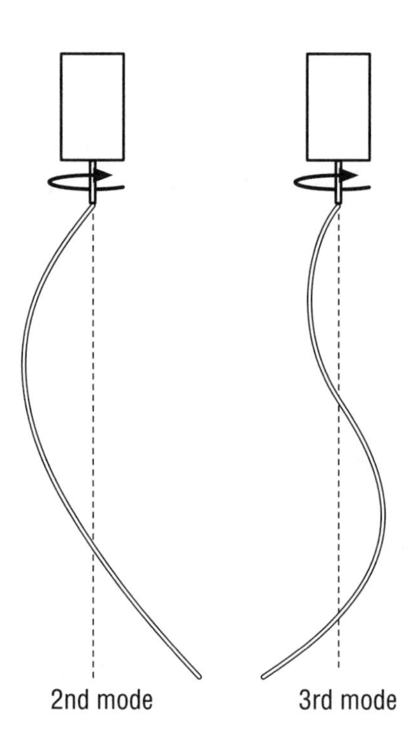

taken slightly off the vertical. It is also noticeable that the rope tends to start off with a higher number of nodes that decrease as equilibrium is reached.

The Helicoseir can obviously be modified by the addition of a simple weight at the bottom: in the limit of a large weight, the modes approach the simple harmonic vibration modes, with the weight defining the tension and thus the speed of the waves.

The mathematics of the Helicoseir, even at equilibrium, are complicated; a detailed description of them appears in Lipscombe, Silverman, and Strange, "String Theory."

REFERENCES

Banks, Robert. *Towing Icebergs, Falling Dominoes.* Princeton, N.J.: Princeton University Press, 1998, 171–178.

Braddick, H.J.J. *Vibrations and Waves.* New York: McGraw-Hill, 1965.

Elmore, William C., and Mark A. Heald. *Physics of Waves.* New York: McGraw-Hill, 1969.

Jones, D.R.H. *Engineering Materials 3: Materials Failure Analysis.* Oxford: Pergamon Press, 1993.

Lipscombe, T. C., M. P. Silverman, and Wayne Strange. "'String Theory': Equilibrium Configurations of a Helicoseir." *Eur. J. Phys.* (1998): 379–387.

Strong String Things

6 *String Nutcracker*

Hence no force however great
can stretch a cord however fine
into an horizontal line
which is accurately straight:
there will always be a bending downwards.

—William Whewell,
"Elementary Treatise on Mechanics"

William Whewell observed, in a famously accidental poem, that a cord stretched out horizontally will always be bent by the force of gravity. Even when that gravitational force is small compared to the force stretching the cord, the cord will form a curve with a slight catenary slope downward. Here we use the inverse of the same cord-tensioning phenomenon to crack nuts.

Boy Scouts and Girl Scouts used to be renowned for the amazing things they could do with a piece of string or rope. However, I doubt whether many even among these seasoned outdoor enthusiasts can crack nuts with a piece of rope. It is a little-known fact—unless you are familiar with large boats—that enormous forces can be generated in ropes by "springing." I am no circus strongman, but I have moved a boat weighing twenty tons hard against its fenders at the dock by pulling on a rope in the right way. It is a simple matter, therefore, to build a device that is capable of cracking nuts (yes, even those really tough almonds) with just finger pressure and a piece of thin rope or wire.

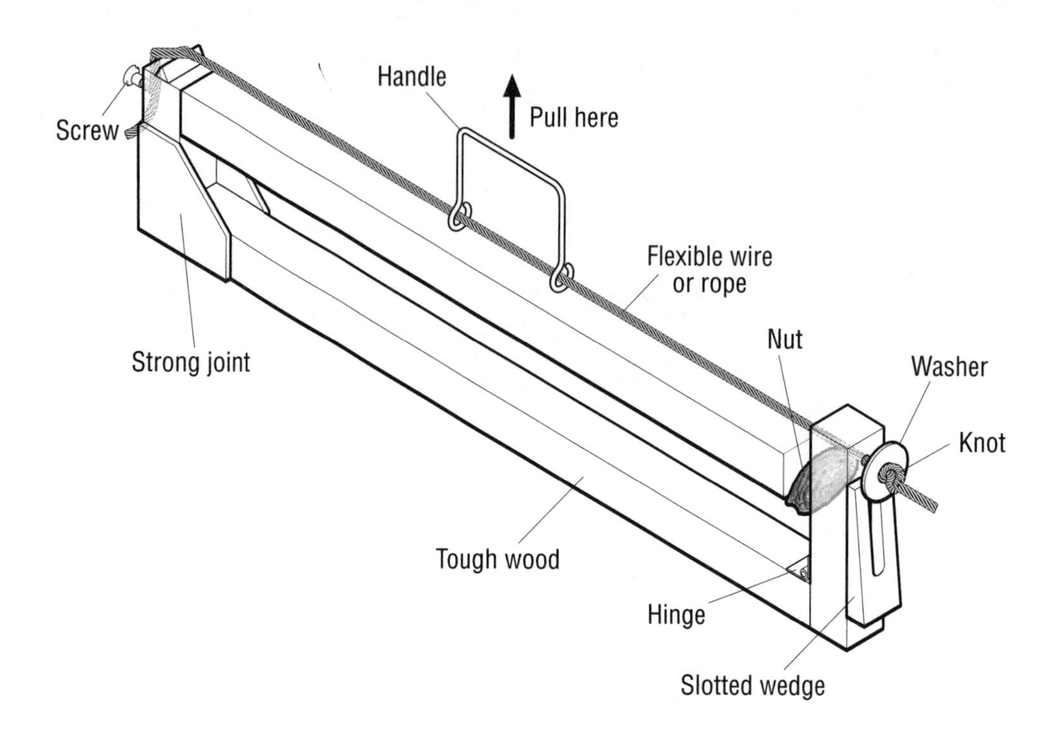

What You Need

- ❏ Wood for frame, about 1.8 m (6 ft) 45 mm (1.5″) square (see figure above)
- ❏ Wood wedge
- ❏ Plywood corner bracing for wood
- ❏ Wire or rope (for instance, 1.4 mm stranded steel or 5 mm nylon cord)
- ❏ Washers to fit on wire or rope

What You Do

Construct a frame as in the figure, and fit and tie off the rope or wire. (Obviously there are many possible variations on the design, but you basically need a pair of jaws to operate on the nut, and a long stick to allow the long rope or wire somewhere to be sprung from.) Where the rope or wire passes through the frame and is then knotted, it will need a metal washer to keep it from pulling through.

The nut is placed in the jaws and the wedge piece used to take the slack out of the jaws. Surprisingly light pressure is then sufficient to crack the nut, which can be broken carefully rather than squashed into lots of tiny pieces that get lost in the carpet! (Simple nutcrackers of the pliers type have the defect that they don't stop once they have cracked the shell, but go on to squash the whole nut—shell and kernel.)

THE SCIENCE AND THE MATH

As you pull the rope, you move its middle a long way, while the ends move only a tiny distance. The geometry of this is shown in the diagram that follows. For small angles α, the angle at the end is the same as the angle formed by the rope, and this is approximately given (in radians) by D/L. The movement of the end against the nut to be cracked is given by $H - L$, and

$$H - L = D \sin \alpha \sim D \alpha,$$

that is, $H - L = D^2/L$.

All these figures (except D) need to be doubled to allow for the fact that there are two ends to the rope, this equation applying to both of them, hence the nut-cracking distance X moved is

$$X = 2D^2/L.$$

(You can get the same result by using Pythagoras's theorem, of course.) For example, with the 400 mm (L = 200 mm) nutcracker, if you move the middle of the rope by 20 mm, the ends move only 1 mm. Finally, we note that there is conservation of energy in the system: none of the energy the user has put in has gone into anything other than cracking the nut. If the force applied by the user at D is F_u and the force applied to the nut is F_n, then

$$F_u D = F_n X$$

as the product of force and distance gives the energy put in by the user. Hence the force applied to the nut is given by

$$F_n = F_u D/X$$

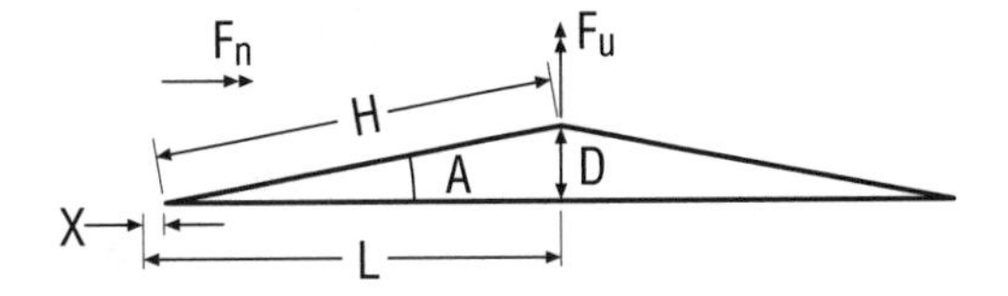

$$F_n = \tfrac{1}{2} F_u L/D,$$

so a force of 50 Newtons (easily given by finger pressure) will apply a force of 500 Newtons—the same as a 50 kg weight—to the nut.

This formula emphasizes that the force magnification effect diminishes hyperbolically with the deflection used. This is why the wedge or a similar mechanism must be used to minimize the gap in the jaws of the nutcracker. If the jaws have to close up before contacting the nut, D will become larger, and the force applied will be smaller for a given user force.

The rope, of course, must withstand in tension the full force that is applied in compressing the nut to crack it. So the rope must withstand, in the example just given, 500 N stress, not just the 50 N applied by the user.

There is also a need to use wire or rope of high Young's modulus. Imagine that you have used a rope of rubber and are comparing its performance with that of a steel wire of the same dimensions. The steel will stretch only minimally (in our example, with a force of 500 N, only by 0.25 mm for a 3 mm^2 cross section) and will thus behave almost like two rigid rods in the geometrical analysis, and the force magnification will be as given. Now consider the rubber rope. It might stretch an additional 1 mm under an added applied force of 0.2 N. Even with a force magnification of 100x, the force will be only 20 N, and this is still not enough to crack a nut. And there is no point in pulling harder at the rubber, because the force magnification will go down hyperbolically, as we have seen.

And for Advanced Users

The wedge in this simple design is a little clumsy and restricts the range of nut sizes that can be accommodated. I am sure that you will think up your own improvements, but here is one you might like to try. Instead of tying a knot in the rope, simply loop it around a large peg fitted into the frame. To operate this version of the String Nutcracker, simply wrap the rope around the peg a few times, holding the loose end: you will find that only a light force is necessary at the end of the rope to keep it from slipping while you pull the length of rope in the middle, as before, to crack the nut. A similar mechanism could perhaps be used with a steel wire too.

Another version of the nutcracker that makes it a smaller unit to store has been suggested to me. Use a table top as the main member, or base, clamping the jaws to the top and looping the actuating rope around a table leg. In this way, only the small jaw-and-clamp assembly and coil of rope need to be stored.

Yet another possibility is often described in books on mechanisms, such as *Kinematics and Dynamics of Machinery* by Walter J. Michel, J. Peter Sadler, and Charles E. Wilson (pages 182–183): use the same geometry but pushing rather than pulling, what is often dubbed a "toggle link." In this, two bars that almost form a straight line are pushed toward straight by a modest force, exerting a large force on their ends if they are confined.

And Finally . . . the Ultimate Nutcracker

You may have already surmised that if one rope-springing stage gives you a force amplification of 10, then two stages could give you an amplification of 100.

The set-up that you might try is given in the accompanying figure. With the input wire moved, say, 100 mm (on a 2,000 mm length), the intermediate wire will move 10 mm (on a 200 mm length), and the final vice grip about 1 mm, while forces will increase from an applied 50 N (5 kg force) to an intermediate 500 N to an ultimate 5,000 N (half a ton!).

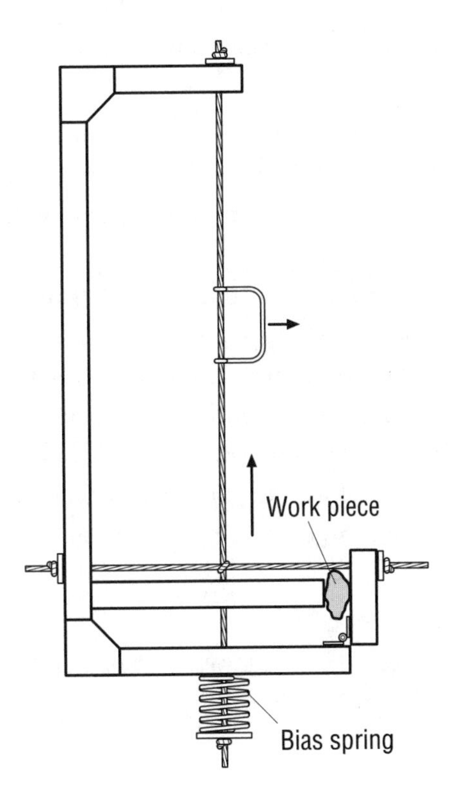

The input stage can be simply made out of wood and a modest thickness of rigid wire. Naturally, the output-stage wire, frame, and jaws must be capable of sustaining the large forces involved and must be very rigid (see the earlier comments on the use of elastic string), which probably means metal construction and a thick wire. The spring is used to provide a "bias tension" to the input stage, just to keep the input rope or wire W_1 straight and taut, and the intermediate wire W_2 straight. For the output stage, this bias tension is provided by the work piece (here, the tough almond nut). But the bias tension on the input stage would tend to close the output jaws when you were trying to load the work piece. The screw is not used to crack the nut. It is simply tightened up finger tight to hold the work piece and to provide a little bias tension for the output stage.

The question arises as to whether the two-stage arrangement offers anything more than could be offered by a single-stage set-up. An argument of the Achilles and the Tortoise type (thoroughly convincing but completely wrong) would probably run along these lines: you can get a force multiplication of 100 directly from a single stage, and you spring a 2,000 mm wire out by only 10 mm. Why mess around with another stage when you've already done it? Or have you?

Some more reflection will probably convince you that two stages are likely to be better:

- A single stage would have to use a very rigid and strong metal stage for the whole device, along with heavy gauge wire.
- With a deflection of 100 mm and a gearing factor of 100, you would need to use a single stage 20 m (60 ft) long!

The argument for using a single stage of 2,000 mm length breaks down because the deflection needed is 50 mm, not the 10 mm of the single stage suggested—with only 10 mm deflection, the output jaw deflection would be only 0.1 mm—not enough to crack even a rather brittle nut.

Of course, you don't need to stop at two stages: with three stages, you could offer an even larger jaw travel with the same magnification, with the same maximum 2,000 mm stage, or a larger force magnification, and with four stages . . .

REFERENCE

Michel, Walter J., J. Peter Sadler, and Charles E. Wilson. *Kinematics and Dynamics of Machinery.* New York: HarperCollins, 1983.

7 *Twisted Sinews*

Tyger! Tyger! burning bright
In the forests of the night
What immortal hand or eye
Could frame thy fearful symmetry?

What the hand dare seize the fire?
And what shoulder, and what art,
Could twist the sinews of thy heart?
And when thy heart began to beat,
What dread hand? and What dread feet?

—William Blake, "The Tyger"

The attachment of muscles to bones and tendons so that they may move a limb has been a source of fascination ever since the work of early scientists described it. The extraordinary Leonardo da Vinci, being at once scientist, engineer, and artist, studied muscles and limb movement in detail. His sketches make it clear that he dissected animals, especially horses, as well as human bodies before painting his masterpieces. Great life artists and sculptors ever since have all studied what is beneath the surface of life.

Two hundred years after Leonardo, the scientific study of animal movement was well under way. Vesalius had expanded anatomical studies, and G. A. Borelli wrote the classic *De Motu Animalium* (On the Motion of Animals) in 1680.

Simulation models, both to illustrate how animals work and to provide practical devices, have been made from those early times, often somewhat gruesomely built on animal or human bones.

However, the simulation of animal or human muscles is rendered difficult by the problem of finding a simple controllable transducer that will exert quite large forces over comparatively small distances. As Steven Vogel notes in *Life's Devices,* muscles can contract in about 200 ms (page 262), although there are muscles ten times faster than this, and ideally a simulated muscle would match this as well as the 200 W/kg power that muscles can produce. Even in models that are not

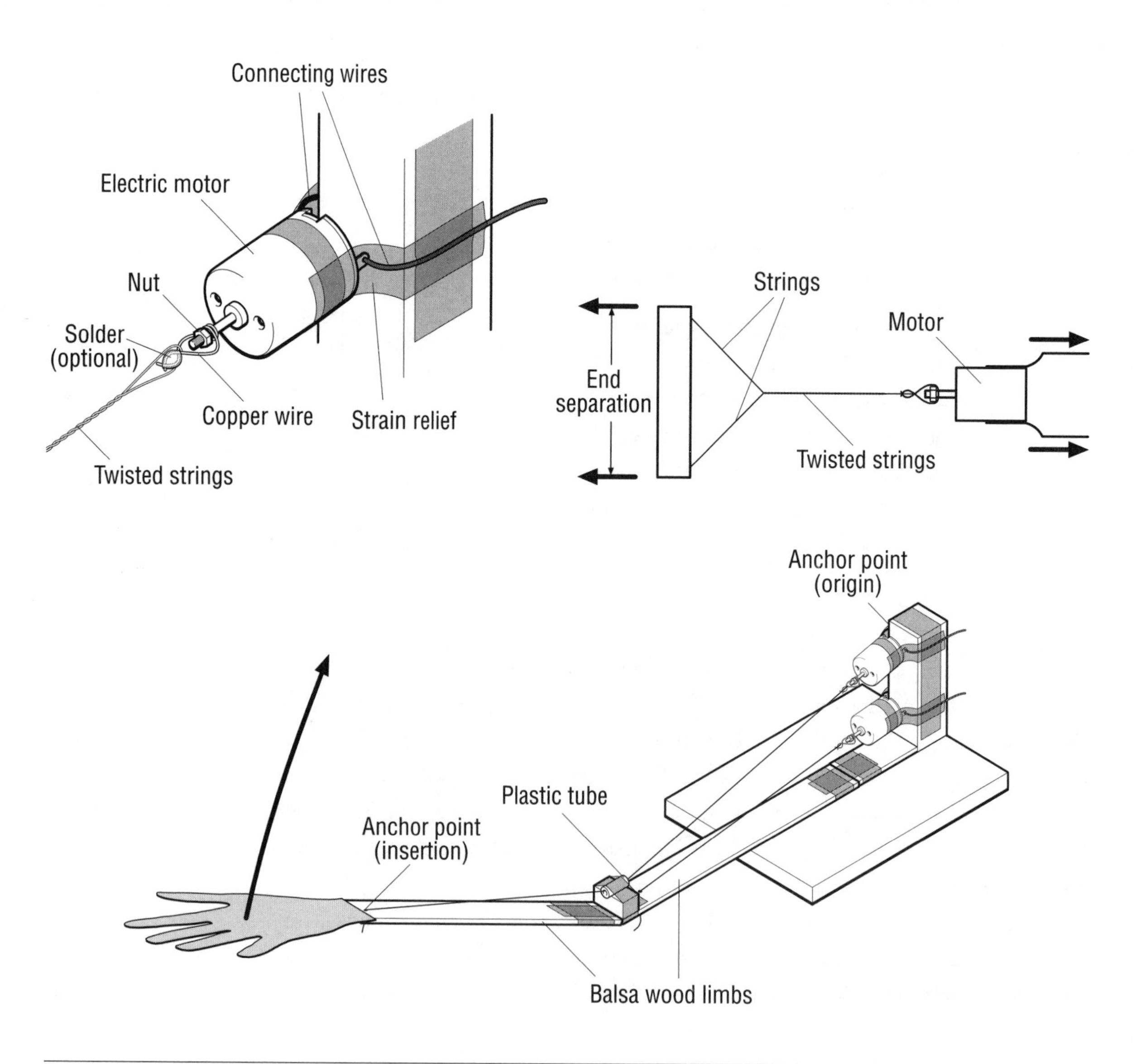

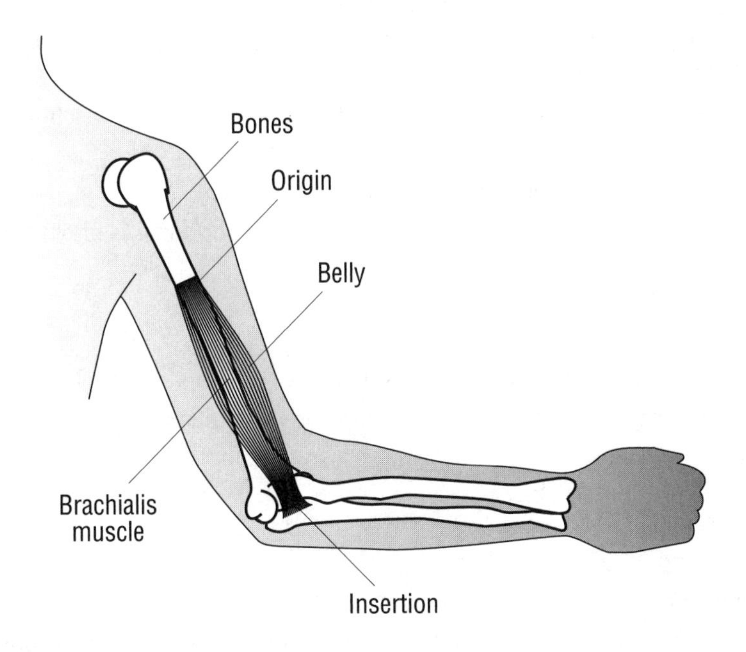

required to operate at natural speed, there is still a requirement to match the force produced by a muscle. Hydraulics are expensive, as are pneumatics, and pneumatics are difficult to make operate smoothly. The more obvious and inexpensive method—electric drive—is not easy either. An electric motor with a high-reduction transmission linked to a screw jack of some kind is often used, but this kind of subassembly is quite difficult to get working and needs many moving parts.

Here we use the motor simply to twist a pair of cords, achieving a smooth and powerful actuating force from the simplest possible electric motor drive.

The Degree of Difficulty

Compared to the other ways of animating an anatomical model, this must be the easiest conceivable. There are few real difficulties, although you do need to ensure that the motor and battery combination you use will not overheat in the stalled (stationary) state.

What You Need

- ❑ Small electric motors (number to match the number of muscles you want)
- ❑ Cord or string of high quality (such as kite cord)

- ❏ A few centimeters of copper wire from solid-core cable (such as 0.7–1 mm diameter or #18–22 AWG)
- ❏ Balsa wood for limbs (for example, in sheets 75 mm × 1 m × 3 mm thick)
- ❏ Cardboard (or balsa wood) for hand
- ❏ Reinforced tape or plastic packaging tape for joints
- ❏ Plastic tubing (outside diameter about 6 mm) or short pieces of old ballpoint bodies
- ❏ Sandpaper
- ❏ Glue gun and glue
- ❏ Small metal nut, superglue, 3 cm stiff copper wire (optional)
- ❏ Springs or rubber bands to form return springs (optional)

What You Do

Roughen the motor shaft slightly with sandpaper. You can simply glue the string onto the shaft with a blob of hot glue from a glue gun. However, a more sophisticated approach will allow easy replacement of string to make comparative tests or to replace a broken string. First, glue a small metal nut (a reasonably close fit to the shaft) to the shaft with superglue. Next, wrap a small piece of stiff copper wire (just 3 cm long should be okay) around the shaft below the nut. Then twist the two ends together firmly, using a small blob of solder to give a stronger joint if this seems necessary, and a further drop of superglue to hold the wire firmly onto the nut.

Secure strong tape to the two connecting wires of the motor and to the motor body to act as a strain relief, so that the weak soldered joint does not take all the tension, and then form the static anchorage (called in biology "origin") of the motor. The two wires must resist the small amount of torque applied when the motor runs. The moving anchorage (the biological term is "insertion") is formed by the distant ends of the twisted string pair, which is tied to the limb at a carefully selected point. The string pair twists and shortens as the motor turns, flexing the limb. With this as a basis, working models of single muscles (like the brachialis shown) can be easily constructed, and multiple muscle models using several motors and various levers and joints are not difficult.

The balsa and tape are used to create the bones and joints, while the tubing is used to feed string pairs through to the second bone of a limb. If arranged as in the last of the series of diagrams shown, then gravity pulls the limb downward, but elastic bands can be arranged to oppose the tension in the twisted-string actuators. Using elastic instead of gravity for the "antagonistic" muscle works better and illustrates animal muscle better.

The Tricky Parts

Nature probably had some difficulty in finding the best compromise among possible places to attach the various muscles to the bones on animals and humans. As the designer of a limb simulator, you too now have this problem. If you attach the strings close to the hinges, you will find rapid movement is possible but force weak; attaching the strings farther from the hinges reduces speed of movement but increases the available force. Also, you will find that, just as in your own limbs, the amount of force available at the end of a limb varies a lot with the angle of the limb. (Incidentally, this is one of the reasons why, when lifting a heavy weight, we are advised to squat next to it and lift vertically, rather than lean over the weight: with the squat lift the angles of the limbs are much more favorable. The other reason has to do with avoiding spinal injury.)

The Surprising Parts

Because the string actuators impose little friction or inertia effects on the electric motors that drive them, the motors used can accelerate almost as rapidly as they do when rotating freely, giving a surprisingly fast speed of response. Although an arm constructed with Twisted Sinew muscles would probably find playing Chopin's "Minute Waltz" beyond its capabilities, the simplicity and low cost of the motor units are such that an arm with fingers is probably not beyond a few hours' work and a pocket-money budget.

I found that for simplicity, hinged joints can be most easily made using cloth-reinforced tape or pieces of package-strapping tape, although actual hinges will clearly last longer. The pieces of plastic tubing, about ½″ long, can be taped to joints to confine the strings to the correct part of the limbs, using small pieces of balsa wood to space them away from the joint to give them a suitable amount of leverage.

THE SCIENCE AND THE MATH

The simplest theory is that the linear distance the two strings span S is reduced by the amount of string needed to allow one string to spiral around the other. For calculation purposes, visualize a thin string wrapped around a thick column in helical fashion. Now further imagine that the string is the hypotenuse (longest side) of a right-angled triangle of paper that has been wrapped around the column. The length of the helix path of the string is now calculated by unwrapping the paper, yielding of course a right-angled triangle with sides L, S, and $N\pi d$ in length. Pythagoras's theorem can be applied:

$$L^2 = S^2 + (N\pi d)^2.$$

Hence $S = \sqrt{L^2 - (N\pi d)^2}$,

where L is the length of each string, d its diameter, and N the number of twists in the string.

This is a steepening curve in which at small numbers of twists N a single twist makes little difference to S, but at larger N a single twist makes a much bigger difference. The steepening curve will be familiar to anyone who has ever wound up a rubber-band-powered toy. In the case of rubber, the length S does not change—the rubber simply stretches—but the tension in the band does change. As the curve steepens, the tension between the two ends of the twisted rubber band increases very rapidly with the number of twists, until a point is reached where either the rubber band or the toy breaks.

This simple theory certainly gives the right sort of curve—although I have adjusted the "diameter" of the string to make the curve fit, rather than using the measured string diameter.

The theory is complicated by a small "chaotic" effect, an effect that is not completely reproducible from trial to trial: the knotting of the twisted string. By "knotting," I here mean the formation of tightly bunched helical turns in the rope formed by the two strings. The onset of knotting was not, in the strings I tried, at precisely the same number of turns each time—it varied by a few turns in 150 or so, a few percent variation. When a "knot" forms, the length spanned by the strings changes slightly more rapidly than predicted by the simple theory.

The onset of knotting is of some importance to the practical application of a twisted-string actuator, partly because knotting is not so predictable, but more seriously because knots can pull out again under tension: the twisted string can behave like a cable made from a "plastic" material, one that deforms under load without springing back. This makes precise positioning difficult. The untwisting of knots when a twisted string is unwound adds another complication. In untwisting, the knots tend to persist a long way past the point at which they formed in the twisting process, an effect that might be dubbed "hysteresis." Finally, there will come a point when the knots on the string occupy all the spanned length, and the "rope of knots" itself begins to knot, giving further complications.

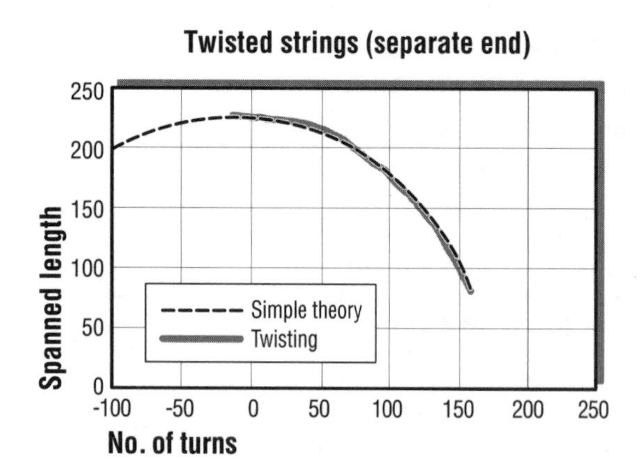

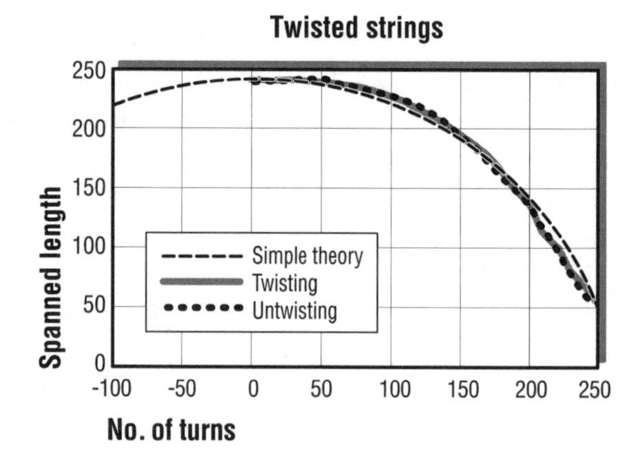

The area where the simple theory curve slopes more steeply near the untwisted state in the graph is genuine. In other measurements too I found evidence for this surprising flat portion of the graph near the untwisted state. Instead of following the parabolic path—which is admittedly very shallow in its slope in the first fifty turns or so with the string used for the graph—the string actually seems not to contract at all with twisting for the first few tens of turns. This could be due to the twisted strings tending to "crimp"—to take on a natural shape—into a slightly helical form once they have been twisted strongly. When the number of turns is zero, this slightly helical shape prevents them from straightening completely to their full length *S*, explaining the flat portion of the graph.

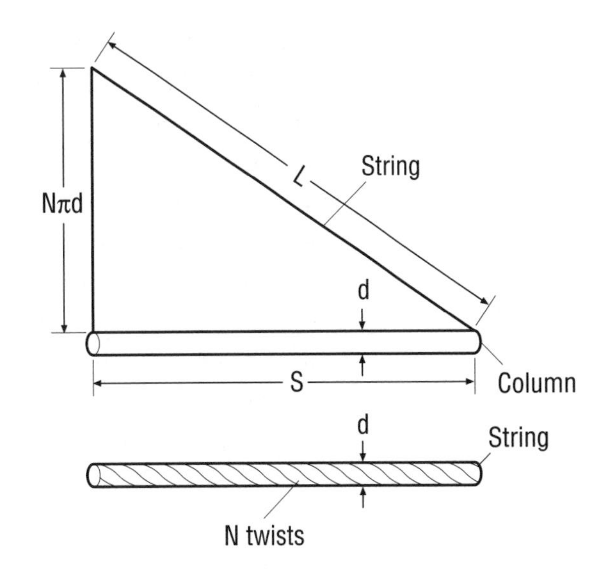

In practice, a twisted-sinew actuator is not very useful in its first few turns anyway—the response is too slow, so this may not be important in practice. This response can be improved by separating the two ends of the string a short distance apart (at the "insertion" end, the end away from the motor). This does modify the behavior of the strings at smaller numbers of turns so that the spanned length *S* is changed relatively more at small *N* than is the case without separating the ends. The improvement can be seen on the second graph, showing theory and measurement.

And Finally, for Advanced Users

The lifetime of the string-twisting mechanisms is limited and varies widely with string type and fiber type. There are immensely strong and abrasion-resistant yarns now available made from manufactured fibers such as polyester or nylon that would last longer than string made from natural materials. Nylon monofilament of the kind sold in fishing tackle stores may show some advantages, although it may suffer plastic distortion if twisted strongly (permanent distortion much worse than the crimping effect described earlier), rather than behaving elastically (that is, not permanently distorting), as happens with conventional string made from many fine fibers.

It is worth noting, however, that the actual strength of the materials is not the only consideration: materials of a high Young's modulus are desirable, as the actuator will then give more precise positioning. To take a ridiculous example, a rubber strip that has a spectacularly low Young's modulus (say, 0.02 GPa) would be useless. Metal wires stretch much less easily, having high Young's modulus, often 100 GPa. Plastics have moduli as low as 1 GPa, although nylon has up to 3 GPa, and other materials, especially those with special processing, higher still.

Reversibility effects are not of course seen in the nonreversible screwed-rod type of actuator, which is also an unrealistically stiff model of a muscular system. The twisted-string actuator is reversible, and even when highly intensible string like Kevlar fiber is used, it will present a realistic flexibility. When, for example, a substantial load is placed in the hand of the arm simulator, the motors will run in reverse when the motor current is switched off. To maintain a particular position, then, a constant smaller current is needed to keep the weight lifted—somewhat like a real muscle.

The reversibility of the actuator is perhaps more obvious when the ends of the string are not next to each other but slightly separated. As noted in the math section, there is an advantage to this, in that the very slow actuation near zero twist that is shown in the graph is avoided: the actuator begins to shorten as soon as it is twisted, albeit still with the steepening curve. With a separation of 32 mm on a 250 mm actuator length, I obtained the curve shown in the accompanying figure, which you can see happens to give a closer agreement to the simple theory curve than is seen in the graphs.

You can separate the strings at both ends, with similar but larger effects. However, in this case, care must be taken not to allow the string to completely untwist, as the motor will not typically have enough torque to put in the first twist in this arrangement.

The low cost of small motors from certain sources (educational suppliers, for example)—often under a dollar—means that you could affordably arrange demonstrations incorporating both extensor and flexor muscles and perhaps joints with more complex movements, such as the wrist. (The extensor is the muscle that straightens a joint; the flexor is the muscle that bends it.) As mentioned earlier, it is difficult to accurately place muscles to give both good speed and force: it is worth studying anatomical drawings of an arm, for example, for ideas as to where muscle attachments might best go.

REFERENCES

Alexander, R. McNeill. *Exploring Biomechanics.* New York: W. H. Freeman and Scientific American, 1992.

Vogel, Steven. *Life's Devices.* Princeton, N.J.: Princeton University Press, 1988.

Strong Nothing

8 *Vacuum Muscles*

Muscle . . . powers the galloping of horses, the flight of bees, and even the swimming of jellyfishes. There is more of this tissue (muscle) than of any other in the bodies of many animals: Muscle accounts for 63 percent of the body weight of trout and 46 percent of the body weight of the Uganda kob antelope.

—R. McNeill Alexander, *Exploring Biomechanics*

Perhaps vacuum actuators—Vacuum Muscles—could be used to improve on nature. After all, with Vacuum Muscles, the trout R. McNeill Alexander mentions in the chapter epigraph would be scarcely a third of its usual weight (which might give it excess buoyancy problems—it would end up light enough to float). And the Uganda antelope would be nearly half its normal weight and able to leap over trees!

Vacuum actuators can be both powerful and fast. Many trains in the past employed vacuum actuators as emergency brakes, and some still do use vacuum (the Westinghouse system was based on vacuum). Train brakes have to be fail-safe, and vacuum-operated cylinders that kept the brake shoes from touching the wheels fulfilled this requirement easily: if there were any leaks or breaks in the brake system, air rushed in and caused the brakes to apply. Automotive engineers have been well aware of the capabilities of vacuum power since powered transport began and have designed a number of gadgets based on vacuum power, mostly on the vacuum derived very simply from the intake manifold of an engine. The suction available from an internal combustion engine is modest (only at best 0.1 atmospheres below atmospheric pressure or less), but it is nevertheless sufficient to operate vacuum servo-hydraulic brakes and air conditioning flap valves, among other things, in today's motor vehicles.

The Degree of Difficulty

Once you have discovered that two- or three-liter carbonated-drink bottles are just the right size to provide the end-caps of 100 mm bellows tubing, you will find Vacuum Muscles quite easy and quick to make.

What You Need

- ❑ 2-liter plastic carbonated-drink bottles
- ❑ Strong fence wire, just a foot or two
- ❑ Bellows tubing, 100 mm (4″) diameter (used to vent tumble dryers)
- ❑ Flexible tubing, 15 mm (outside) diameter (large garden hose, or the ribbed hose used to expel wastewater from household machines)
- ❑ Strong reinforced tape
- ❑ Small hinges (door or furniture)
- ❑ Wood and balsa wood
- ❑ Plastic tubing, 4 or 6 mm diameter (or short pieces of ballpoint bodies, carefully smoothed off inside)
- ❑ Strong wear-resistant string (braided cord is best)
- ❑ Strong rubber bands or bungee cord
- ❑ Vacuum cleaner with hose
- ❑ Small plastic plumbing-pipe T-piece, 15 mm diameter, and large washer

What You Do

Each Vacuum Muscle is produced by cutting the top off two carbonated-drink bottles, about 100 mm down. Seal one bottle with its screw top, and attach the top of the other bottle to the flexible tubing (hose). The hose end can be attached by the hose itself as the origin (fixed end) of the muscle.

Loop a piece of tough fence wire around the neck of the bottle with the top on it. Bring the ends of the wire to opposite sides and twist them together about

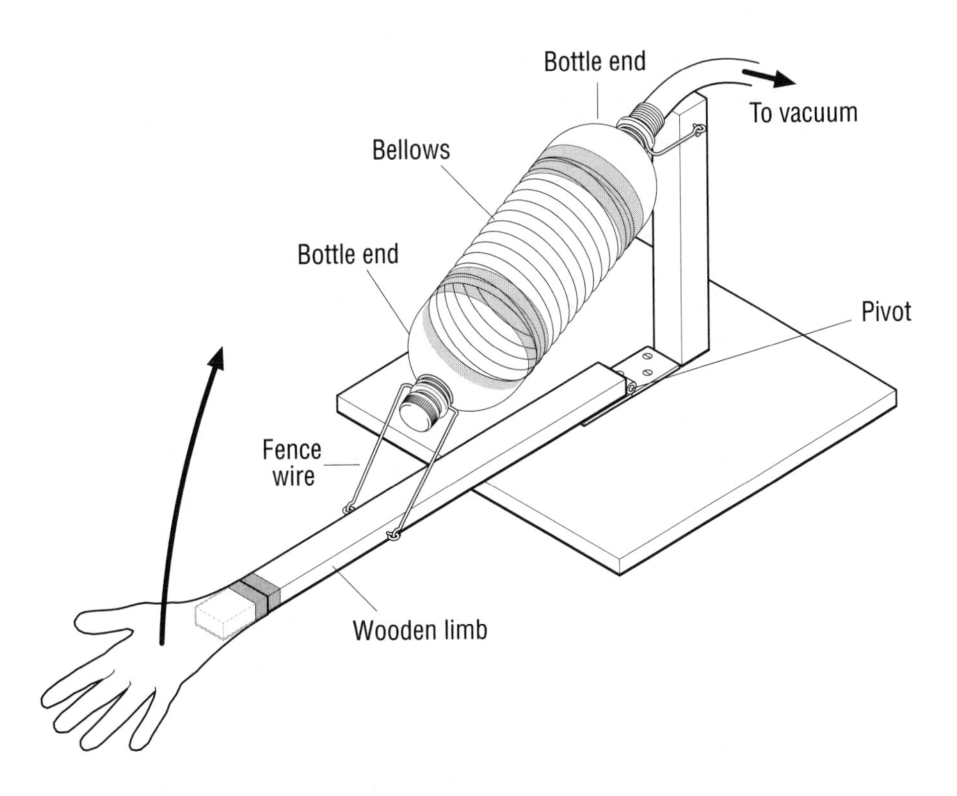

50 mm from the neck. This forms a center pull to attach the muscle to its insertion (moving end) or to attach a cord tendon to. Alternatively, you can use two eyes screwed into either side of the wooden limb in which the two wires from the bottle top pivot, as in the figure above.

You should start with bellows about 100 mm (4″) long or so: this avoids some instability problems, as we will see. Cut the bellows 50 mm (2″) longer than finally required and then insert the bottle ends inside the bellows. Tape smoothed around the joint completes the job. The origin and insertion should be pivoted or attached by string, to keep the bellows straight as it contracts. The bellows tubing comprises a thin jacket of very flexible plastic (perhaps plasticized PVC) over a helical support of piano wire (a very strong spring steel). Because of its spiral wire support, the bellows twists slightly as it contracts, and this should be allowed for. However, despite these minor problems, these Vacuum Muscles still can easily form the basis of good models of limb movement.

As with the earlier Twisted Sinews project, the limbs for the Vacuum Muscles can be pieces of balsa wood and the hinged joints can be made with tape. However, with the greater forces here, metal hinges are probably advisable. For the same reason, I used pine wood for the limbs, because it is stronger and it is easier to screw hinges to it. Once again, pieces of plastic tubing about ½″ long can be taped to the joints to confine the tendons to the correct part of the limbs.

If the Vacuum Muscle itself is acting as a flexor muscle, which bends a joint, then the bungee cord or heavy rubber bands can be used to create the extensor muscle, which straightens a joint, rather than using gravity, as shown in the diagram. The bungee cord can be tied between suitably placed screws.

The vacuum cleaner should be connected to the hose—for instance by connecting or inserting the hose into a crevice-cleaning device attached to the vacuum cleaner.

If your vacuum cleaner works well, you may find that the negative pressure produced even with the cleaner turned to low speed may be too powerful for the bellows. In this case, you can regulate the vacuum to some extent by the leak tightness of the joint with the crevice tool on the vacuum cleaner.

A more sophisticated solution is to make a vacuum divider as shown in the figure below, using the plastic plumbing-pipe T-piece. Air flows in from the atmosphere via an opening to reduce the vacuum achieved at the output, while the rate at which air is removed by the vacuum cleaner is restricted by the other, smaller opening. I covered the pipe opening that attaches to the vacuum cleaner with a small plastic cap, in which I made a hole; I taped a large washer over the pipe opening on the atmosphere side. This is the pneumatic analogue of the electrical potential divider, in which the electrical resistors have been replaced with openings.

The Tricky Parts

Again, as with the Twisted Sinews limb simulator, you will probably have some difficulty deciding how to attach various muscles to the bones. With the

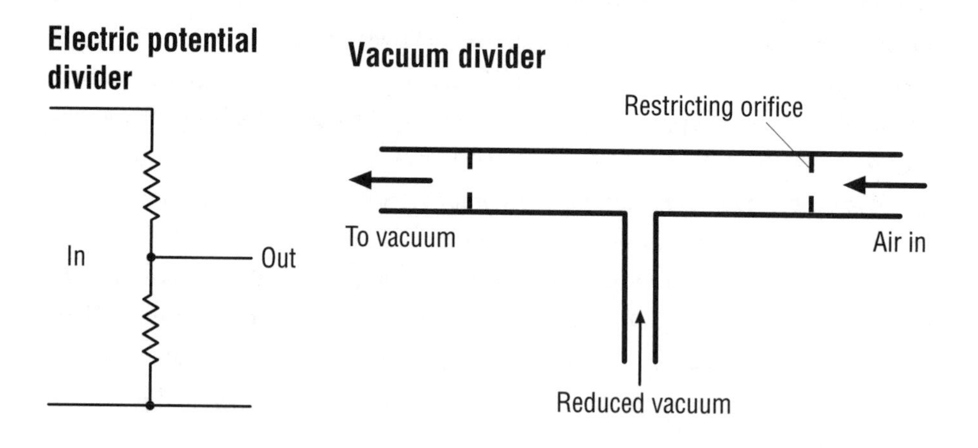

greater power available from the Vacuum Muscles, however, this will be less of a problem.

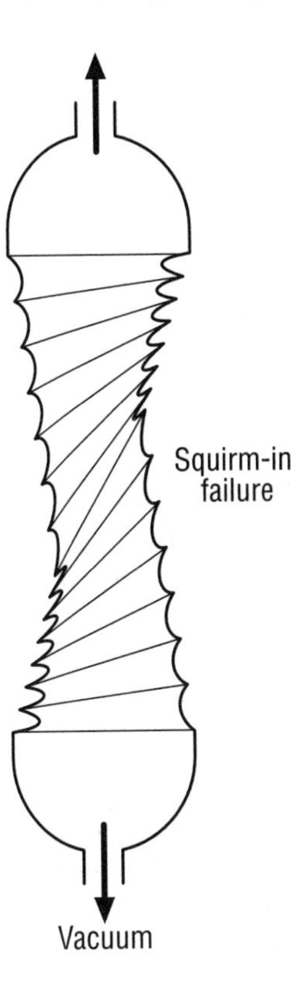

A more difficult problem is not only how to connect and disconnect the vacuum to tense the muscle, but also how to vent the atmosphere into the vacuum chamber when the muscle is relaxing. With a single Vacuum Muscle operating, simply connect or disconnect the tube from the vacuum cleaner. When the disconnection is made, air will vent into the bellows automatically. With several muscles, some of which must operate simultaneously, this is not satisfactory. If a simple on/off valve is used, air will not vent back into the bellows when the vacuum is shut off. A possible solution is a small vent hole on the bellows side of the pipe that is not big enough to have much effect on the vacuum achieved, but this gives only a limited relaxation (muscle extension) rate, since the vent hole must be small.

If you need to use a long bellows section (say 200 mm) in order to get a vacuum muscle with a long travel, then you may have problems. When you apply a high load and high vacuum to a long bellows, you will discover that there is a failure mode in which the bellows wire squirms sideways as the vacuum is applied. (I call this "squirm in" to avoid confusion with a related instability when applying positive pressure.) The helical wire support is usually permanently bent into a leaf-shaped section (two arcs of circles joined by sharp bends) when this happens, and the Vacuum Muscle will subsequently squirm in at a smaller vacuum level.

The Surprising Parts

As with the Twisted Sinews project, the absence of any parts with significant friction or inertia means that the Vacuum Muscles can give a surprisingly fast speed of response, provided the vacuum supply pipe and the atmospheric vent are large enough. An arm constructed with Vacuum Muscles probably could be made to play Chopin's "Minute Waltz." (In fact, some church organs once used vacuum actuators, employing small bellows made from chamois leather and wood, to operate valves at the bottom of pipes that could not be placed conveniently near the console of the instrument. A set of small-diameter tubes led from the console to the vacuum actuators. Even today, pneumatic-action pipes are still occasionally used, with small bellows operating under a small positive pressure.)

The Vacuum Muscle also combines its speed with quite considerable power: a pull of 20 kg is perfectly achievable, and my prototype arm was fast enough to be able to throw tennis balls and even lightweight basketballs. I made a very serviceable hand, by the way, by stuffing a rubber glove with small polystyrene beads coated with a little two-part glue to make them stick together. Hold the hand in the shape you want until the glue sets.

THE SCIENCE AND THE MATH

The maximum force F_{max} that the actuator can exert is simply given by the area A of the end-caps of the bellows multiplied by the vacuum pressure ΔP (atmospheric pressure P_a minus the absolute pressure P) that is applied:

$$\Delta P = P_a - P$$

$$F_{max} = A\Delta P.$$

Its speed of movement is limited by airflow considerations. However, with the high pumping-rate capacity of the vacuum cleaner powering it, only the pipe down which the air is sucked and vented needs to be considered. The smooth flow of oil in a narrow pipe is qualitatively different from the turbulent buffeting flow of air inside a large pipe at high flow rates. The flow in the former case (laminar flow) is a simple formula, but the case we have here is turbulent flow, as can be gauged by calculating the Reynolds number.

$$\mathrm{Re} = \rho Vr/\mu,$$

where ρ is the density, V the mean velocity, r a characteristic length scale (taken to be the radius here, as the tube is relatively long), and μ the viscosity. With Reynolds numbers in excess of 2,000 (I calculated Re = 10,000 for my system), turbulent flow will tend to occur, particularly with rough-walled tubing in short lengths. For such turbulent flow, an approximate rule (Darcy's rule) is that the flow velocity V will be given by

$$V \sim k\sqrt{\Delta P/L\rho r},$$

where k is a constant and L the length of the pipe. This gives, for the volume flow rate Q:

$$Q \sim k'\sqrt{r^3\Delta P/L\rho},$$

where k' is constant. It is interesting to see that the length of the tube is not all that important: if you double its length, you will lose only 30 percent in flow rate. Similarly, increasing the pressure available is relatively ineffective. However, the exponent of 1.5 in radius (square root of cube factor) says that you should always use the maximum possible tubing diameter. Interestingly, if the flow were laminar flow, the factor in the tube radius would be even higher, radius to the fourth power.

Because of its importance in industry, the buckling of tubes and pipes under external pressure has been extensively studied. Unfortunately, it is also complicated, and it is often calculated by computer these days.

C. R. Calladine, in the *Theory of Shell Structures,* has given a formula that describes most of the simpler situations, such as a thin-walled cylinder, for example:

$$P_{cr} = \frac{0.86E(t/L)(t/r)^{1.5}}{(1-u^2)^{0.75}},$$

where P_{cr} is the critical pressure that will cause buckling, and the parameter u a constant of the material, the Poisson ratio. This equation applies to the middle region shown in the graph, which also shows the likely value of the number of lobes in the buckled component.

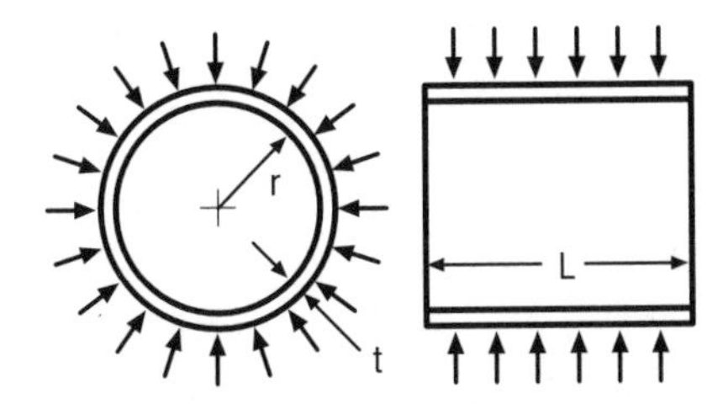

Tube ends held circular

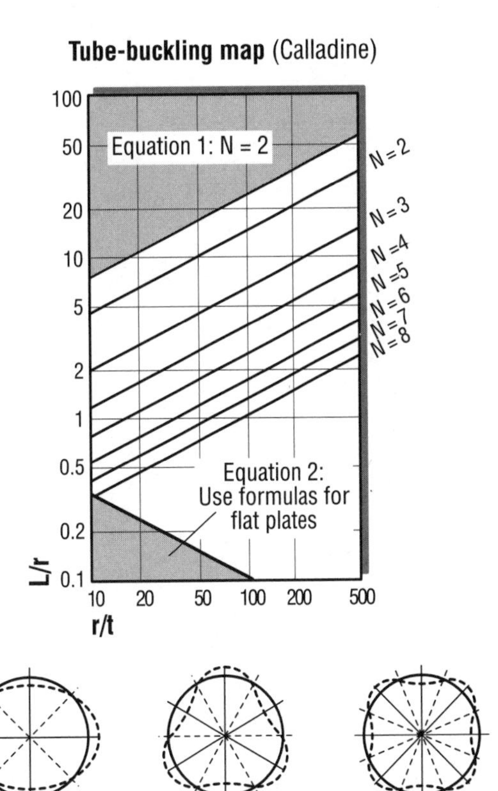

N = 2 N = 3 N = 4

The Poisson ratio is a property of materials. It is the ratio of percentage shrinkage in width of a material divided by the percentage extension in length when the material is put under tension. It lies between about 0.2 and 0.4 for ordinary solids (highly anisotropic materials such as glass-fiber-reinforced plastic can have very low values). As can be seen, the form of the equations and the number of folds formed in the buckled part after failure depend on the two ratios *L/r* (length/radius) and *r/t* (radius/thickness). For the upper range shown in the diagram, another simpler equation applies, while for the lower range, the same equation applies as for flat plates. The bands indicate how many lobes the buckled cylinder will form.

The buckling of the more complex structure of the bellows is naturally more complex, but similar considerations probably apply. The helical wire support certainly is capable of buckling inward into two lobes, or cusps, although I have also seen three lobes form on a bellows that had previously been bent slightly. Sometimes the bottle parts (particularly if you make them too long) will also buckle inward to form three cusps, in the case of the bottles I used.

And Finally, for Advanced Users

The Vacuum Muscle can clearly be used to push as well as pull, at least to some extent. If a source of pressure greater than atmospheric is available, then this can be tried. With short lengths and low pressures, this is okay. However, if the pressure is too high, even with short lengths the plastic covering will tend to belly out from the supporting wire helix and then not go back to its original position on relieving the pressure. After a relatively small number of cycles, this will also lead to failure.

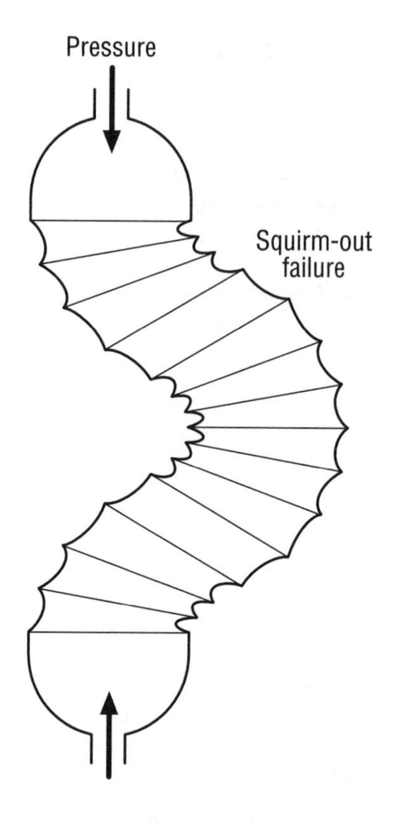

Similarly, even at lower pressures, with longer lengths of bellows things may go awry. The bellows may "squirm out," belly out into a large S-shape, and then (if more pressure is applied) fail. Squirm was responsible for the failure of some steel bellows in eastern England one weekend in June 1974. Unfortunately, as D.R.H. Jones describes in *Engineering Materials 3*, these particular bellows contained hot cyclohexane (a compound similar to ordinary automobile gasoline). A cloud of cyclohexane vapor and air equivalent to 45 tons of TNT detonated disastrously, killing twenty-eight and injuring eighty-nine people, and giving the small village of Flixborough unwanted fame.

As with the twisted-string actuators, with more Vacuum Muscles and valves, you can fairly easily arrange demonstrations incorporating both extensor and flexor muscles. With two opposing muscles, you will be able to hold an arm in position by balancing the vacuum pressure applied to each. Joints with more complex movements are obviously possible too. The speed of the Vacuum Muscles is never a problem: however, if higher vacuums are needed, then short lengths of bellows with short strokes are required to avoid squirm in, and this must be allowed for.

REFERENCES

Alexander, R. McNeill. *Exploring Biomechanics.* New York: W. H. Freeman and Scientific American, 1992.

Calladine, C. R. *Theory of Shell Structures.* Cambridge: Cambridge University Press, 1983.

Jones, D.R.H. *Engineering Materials 3: Materials Failure Analysis.* Oxford: Pergamon Press, 1993. (See esp. p. 346.)

9 *Vacuum Bazooka*

I shoot the Hippopotamus
With bullets made of platinum,
Because if I use leaden ones
His hide is sure to flatten 'em.

—Hilaire Belloc,
"The Hippopotamus"

Curiously enough, platinum makes better bullets than lead does. First, as Hilaire Belloc suggests in "The Hippopotamus," lead is soft and thus does flatten easily. Second, platinum is much more dense than lead (21.5 kg/liter, as opposed to 11.3 kg/liter). This means that, compared to lead bullets, platinum bullets fly farther and have a greater impact on the target when launched at the same muzzle velocity, as they have greater kinetic energy. As the hippopotamus has the reputation of being perhaps the most dangerous animal in the African jungle—more aggressive than a lion or tiger—perhaps platinum bullets are a wise precaution.

Platinum does not seem to have caught on to any great extent, however, because of its enormous cost: it is normally more expensive than gold. In fact, in the shooting of game birds there is a suggestion that lead shot may be replaced by copper or even iron shot (density only 7.9 kg/liter), as the lead pollutes the soil around the gun positions. These iron or copper pellets perform rather poorly except at shorter ranges. There are military bullets made of uranium ("depleted" grade, density 19 kg/liter), since uranium is, surprisingly, rather cheap.* The uranium is not being used for a nuclear weapon, however, but simply because it is heavy!

Rather than using pressurized air or gas, the Vacuum Bazooka generates a vacuum in front of a projectile and uses the pressure of the atmosphere to propel the projectile. However, this gun, unlike the regular kind, would not work very well with platinum projectiles. It prefers low-density bullets (see "The Science and the Math" section for an explanation).

The use of vacuum technology for propulsion is by no means new. Vacuum drive was, rather briefly, the best technology available for railroads. Isambard Kingdom Brunel (1806–1859) and a number of other master engineers of the early 1800s built vacuum (they correctly called them "atmospheric") railroads. A piston running inside a tube pulled the trains, the piston being propelled by atmospheric pressure, the tube in front of the train being evacuated by a vacuum pump miles ahead along the track. Atmospheric air in the tube behind the train pushed the piston, which was connected to the train via a narrow slot. By the standards of the time, vacuum railroads performed well, as they could employ light, high-speed carriages; the necessary steam engines were stationary pumps, which could be large and powerful without increasing the train weight. Conventional trains have to haul along their own steam engine, the "locomotive engine," which must be limited in size, and whose large weight was a big disadvantage when accelerating or when ascending steep gradients.

However, vacuum railroads required a continuous seal along the slot on the top of the vacuum pipe. This seal had to be made of leather and iron, and it was beyond the technology of the time to create a long-lasting seal. (The seal had to open to allow a narrow blade to pass from the piston to the locomotive. The seal then had to close again afterward so that a vacuum could be pulled on the pipe behind to get the train along the line on its return journey.)

D.R.H. Jones in *Engineering Materials 3* describes some of the problems with the seal. Brunel's seal had difficulties in the short term with freezing, and in the longer term was corroded and even eaten by rats. The system, like many, was replaced after a few years, although one vacuum railroad lasted forty-three years in service. Had electricity been slower in coming, and had a suitable rubber or plastic seal been available, maybe today's subway trains would still be using this clean and safe system.

*Depleted uranium is ^{238}U, which can be used in a nuclear reactor only after it has been in a breeder reactor for conversion into plutonium, which can be used to make nuclear power. Because no breeder reactors have been constructed, except for a few pilot plants, the ^{238}U is essentially useless, at least for the next few decades, and its price is thus very low. Natural uranium is only slightly radioactive, and depleted uranium even less so.

The Degree of Difficulty

The demonstration should prove fairly easy to carry out, although it is important to purchase a T-piece that will allow the free passage of the projectile. Some T-pieces have a slightly narrower bore than the pipe they fit, making them unsuitable, while others have ridges or other features that will need to be removed. And obviously, the most powerful vacuum cleaner will give the most powerful bazooka!

What You Need

- ❑ Duct tape or other heavier grade tape
- ❑ Piece of plastic drainpipe and a T-piece, about 35 mm bore
- ❑ Vacuum cleaner
- ❑ Soft wood or balsa wood projectile that fits inside the drainpipe freely but snugly
- ❑ Piece of paper or thin plastic
- ❑ Piece of hard sponge rubber or plastic (perhaps half a sponge ball)
- ❑ Lightweight plastic flaps with tape hinges, instead of paper (optional)

What You Do

Assemble the cross-arm of the T-piece onto the end of the drainpipe as shown in the diagram, making sure that the T-piece presents no obstacles to the free passage of the projectile. (You sometimes find that the T-piece has small tabs molded into its inner surface to locate fitted pipes precisely. If these stick out into the bore, they must be removed.) Then connect the vacuum cleaner hose to the T-piece, using tape to give a reasonable seal. Switch on the vacuum cleaner to suck the air out of the drainpipe.

Partially insert the projectile at the end of the pipe, making sure that you keep a good grip on it. Then take aim with the end of the T-piece (the muzzle), and put a piece of paper over the muzzle. You will now feel the atmosphere pushing hard and you will find it difficult to keep hold of the projectile, indicating a good vacuum. You will hear the note of the vacuum cleaner's electric motor change, and this too will indicate when a good vacuum has been achieved. Let go of the projectile after a second or two . . . whoooooosh!

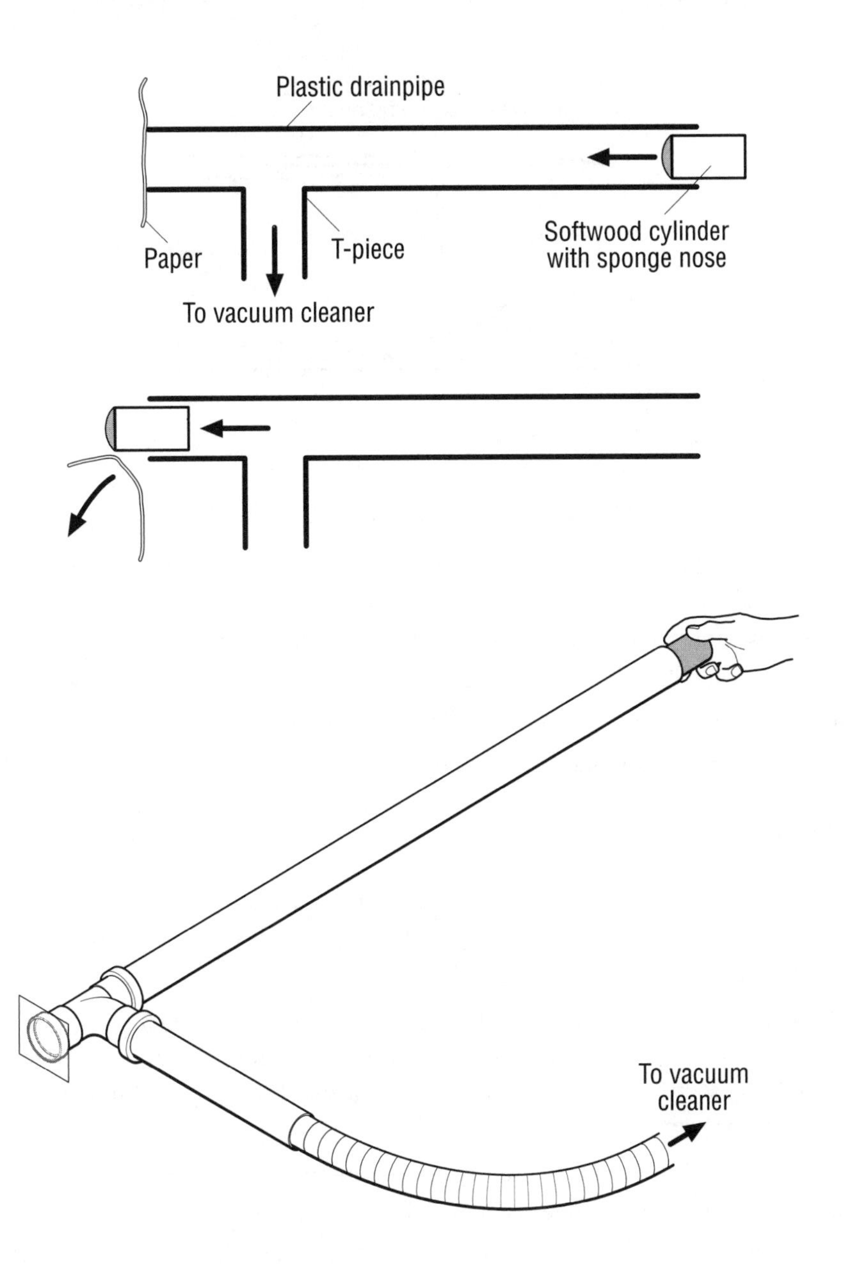

The Tricky Parts

It is tricky to make the projectile fit well enough that you don't lose too much air through leakage, but still loose enough that it goes easily down the tube with little friction. I used a bit of soft pine wood from the leg of a wooden stool, which was just a shade too big, so I sanded it down to size. You could try using a cloth or felt wrapping around a projectile that is slightly too small, or you

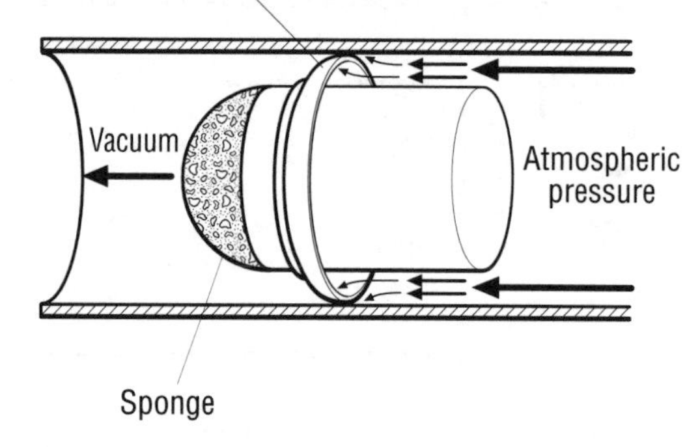

could try making a sealing ring. Sealing rings work by being flexible and cup shaped, so that the pressure, in this case air pressure inside the cup, inflates the cup to make it an air-tight fit in the pipe, as the diagram shows. Bicycle pumps and many industrial seals rely on cup-shaped sealing rings.

The projectile should be long enough that it will not easily jam sideways in the pipe or more especially in the T-piece (although this is rare). A 2:1 aspect ratio is probably about right, so a 35 mm caliber projectile should be 70 mm long.

The Surprising Parts

- When you cover the end of the pipe with the paper, the vacuum pressure suddenly decreases by a factor of a hundred or more, and the projectile is accelerated very rapidly down the tube.
- When the projectile reaches the piece of paper, which is being held on firmly by the air pressure, it doesn't stop but whooshes out into the open air—up to a height of 20 m in the case of one I made (and a long way out of our garden).

The projectile must reach a certain minimum velocity to push the piece of paper off the end—"escape velocity," if you will. With a very small vacuum applied, and with a leaky projectile, you may find that the projectile will simply hit the muzzle paper and stop.

Tip-Top Performance Tips

- The vacuum cleaner might give better suction without a bag or filter in it.
- The projectile may go better if it is better streamlined and equipped with tail fins, but these must be fitted on within the tube diameter.

The Dangerous Parts

Don't leave anything like your antique Ming vase (or your hand, your kid brother, etc.) in front of the muzzle paper after you've put it on. If your grip slips on the projectile, it will shoot up the pipe and could cause an accident. I suggest that you include a piece of fairly hard sponge rubber or sponge plastic—maybe half of a sponge ball—on the nose of the projectile to minimize the damage, should an accident happen.

THE SCIENCE AND THE MATH

The vacuum cleaner produces a vacuum within the tube. The absence of air pressure on the side of the projectile pointing toward the inside does not of itself make the projectile move. What moves it is the pressure (10 N or about 1 kg weight on every square centimeter) of the air around us. (The projectile doesn't move in the absence of the vacuum cleaner's suction because it then has the same air pressure on both sides.) The projectile is not stopped by the piece of paper because, although the paper is capable of exerting a fairly strong force, it will only be able to exert that force momentarily before it is pushed off the end of the tube.

If the vacuum cleaner produced a perfect vacuum (0 atmospheres absolute pressure), then the projectile would move extremely fast. In fact the vacuum cleaner produces a partial vacuum of, say, 80 percent of 1 atmosphere absolute pressure. The projectile rushes down the tube under a pressure difference of 20 percent of an atmosphere, reaching a speed at the end (muzzle velocity) of typically 10 m/s. The paper is of little mass and exerts a restraining force (corresponding to the 20 or 30 percent of an atmosphere pressure) on the projectile for a distance of only a few tens of millimeters. (This assumes that the vacuum cleaner sucks the air away fast enough to ensure a constant pressure during launch.)

$$\text{Force} = \text{Mass} \times \text{Acceleration},$$

$$\text{so that Acceleration} = \text{Force/Mass}.$$

So a projectile of weight 0.1 kg will accelerate at about 100 g, since the force exerted is about 100 N: Force = Pressure × Area, Area = $\pi \times \text{radius}^2$, so, with 0.2 atmospheres of differential pressure,

$$\text{Force} = 0.2 \times 100{,}000 \text{ N} \times \pi \times (40/1{,}000)^2 = 100 \text{ N}.$$

Now the weight of that projectile is simply $g \times 0.1$ N ~ 1 N for a value of the acceleration due to gravity at the earth's surface of $g = 9.8$ m/s^2, and the force we are applying is 100× this much, giving a 100 g acceleration.

The vacuum bazooka will work if scaled up or down in size: if the device is scaled up by a factor of F, then the projectile weight goes up by F^3, while the area over which the atmosphere acts goes up by F^2, and the length over which this force acts increases by F, giving a total projectile energy that scales as the weight scales, and hence a constant muzzle velocity.

Clearly the maximum achievable velocity is limited, however, since atmospheric pressure is only 100,000 N, and even this can only be exerted if the vacuum applied is much better than a vacuum cleaner will typically manage. Equating the work done by the atmosphere on the back of the projectile and the kinetic energy of the projectile, the velocity limit of the vacuum bazooka can be seen to be given by

$$V = \sqrt{2P_0 L/z\rho},$$

where P_0 is atmospheric pressure, L and z the lengths of tube and projectile respectively, and ρ the density of the projectile. In fact, the maximum velocity is even lower than this in practice: if L is made too long, then there are losses due to friction from the projectile sliding and the flow of air from the atmosphere along a long tube. Also, if the projectile density is too low, and the length z is short, the projectile will slow down quickly in the air once it is launched, due to air drag forces.

There is finally the question of "escape velocity." You can make a rough estimate of how big this is by first calculating how much energy the muzzle escape consumes.

Assume, for example, that the end paper is very small in mass but continues to exert a pressure equal to the driving vacuum on the end of the projectile for a distance equal to half the bore, say 17 mm. The amount of E_{abs} energy absorbed by the end paper in this way is thus about

$$E_{abs} = P_0\pi(35/1{,}000)^2/4 \times (17/1{,}000)$$
$$= 0.2 \times 100{,}000\pi(35/1{,}000) \times (17/1{,}000) \sim 0.35\,\text{J}.$$

The kinetic energy of the projectile must clearly exceed this value, that is,

$$\tfrac{1}{2}mV^2 > E_{abs},$$

so that the escape velocity V_{esc} is given by

$$V_{esc} = \sqrt{2E_{abs}/m}.$$

For the 0.1 kg projectile discussed here, "escape velocity" is thus about 2 m/s. The energy absorbed by the end paper is of course absorbed even in the case of an escaping projectile, implying that some percentage of the projectile energy is lost—4 percent in this case.

And Finally, More Advanced Vacuum Bazookas

A subtle question: why does the paper seem to survive most of the time? Why doesn't the projectile always rip it to shreds? Does this have something to do with the fact that the vacuum cleaner only produces a *partial* vacuum? Try taking video camera footage of a launch and see if you can tell what is going on using the freeze-frame facility. Even better, maybe you can rig up a camera to take a flash picture of a launch (perhaps you know someone with a sound trigger, an electronic gadget that sets off the flash and the camera when it detects a loud noise).

In fact the pressure in the drainpipe varies from the maximum partial vacuum at the beginning of launch to a minimum as the projectile compresses the air remaining in the tube. Toward the end of its travel, the projectile is going so fast that the remaining air will typically not be sucked away fast enough by the vacuum cleaner. The pressure builds up sufficiently that right before the projectile emerges, the pressure just in front of it will be positive, and it will blow the paper off the end. This is a benefit, but of course this also means that the projectile will achieve a muzzle velocity lower than it could if vacuum were maintained.

However, by using a large-diameter hose to connect the vacuum cleaner, or perhaps by using a vacuum reservoir, this effect can be reduced and the muzzle velocity increased. The vacuum reservoir might simply be another piece of drain-pipe, perhaps of larger diameter, joined via an adapter to the T-piece to which the vacuum cleaner is connected.

The paper can be replaced by a pair of light flaps, either free or hinged. I tried taping a flat plate with the hole in it over the muzzle, and attaching flaps to that with hinges. If the flaps are heavy, clearly the projectile will tend to bounce off them. For example, if the flaps weigh the same as the projectile, then the projectile will simply be stopped, with the flaps flying off with more or less the velocity that the projectile would have had (what is called in games of snooker, billiards, or pool a "stun shot"). This follows from the laws of the conservation of momentum and energy in mechanics.

This raises the intriguing possibility of replacing the paper with another projectile, but this projectile would need to be prevented by a rim or similar device to stop it from being sucked back into the tube during launch. A tennis ball can be used with good effect and has the advantage of being safer in case it goes astray and hits anyone. The wooden piston (now no longer the projectile) rushes down the tube and whacks the tennis ball à la John McEnroe. If the piston has the same mass as the tennis ball, then almost all of the piston's energy can be transferred to the ball, leaving the piston just about stationary after knocking the ball up in the air. You could even ensure that the piston doesn't leave the vacuum tube by retaining it with a piece of string or elastic. To reload, you simply remove the vacuum for a few seconds while you pull the piston back (or allow it to fall back, under gravity) and load another tennis ball.

However, if the collision between piston and ball is inelastic to some extent, energy is lost. If, for example, the piston stuck to the projectile in the muzzle, then the system would be rather inefficient. With ball and piston stuck together as the same mass, the velocity available would be halved, and the kinetic energy halved too. If the angle T of the flap is narrow, although the projectile will be deflected, the energy lost due to the impact might be reduced relative to a flap orthogonal to the tube.

Measuring the Muzzle Velocity

You might like to think of some good ways to measure the muzzle velocity. You could of course use gated photoelectric cells and electronic counters. However, purely mechanical means—collisions or trajectories, for example—might be more elegant. Firearm muzzle velocities used to be measured by the swing of a massive pendulum into which they were fired, using the law of conservation of momentum, although today electronic techniques are used.

A simpler electronic set-up might be as in the accompanying diagram. The projectile is arranged to break through two pieces of foil (I used two ¼″ wide strips of thin kitchen aluminum foil). The first piece keeps the voltage on the capacitor at ground potential. The second piece of foil connects the capacitor C to a power supply of voltage V_0 via a resistor R. During the short time interval between the projectile breaking the foils, the capacitor is charged by a voltage V given as follows:

$$V = V_0[1 - \exp(-t/RC)].$$

If the resistor and capacitor are chosen so that V is much less than V_0 (say, one-tenth), then a simple linear law can be used, as for small t/RC, $\exp(-t/RC) \sim 1 - t/RC$.

$$V = V_0[1 - (1 - t/RC)] = V_0 t/RC.$$

The capacitor C should be big enough that the digital voltmeter used to measure the voltage across the capacitor will not appreciably discharge it while you are reading it out. Some example values: if the break foils are 0.5 m apart, and the projectile is traveling at 10 m/s – 1, then the time interval is typically 50 m/s. With a 500 microfarad capacitor C and a 1,000 ohm resistor R, readings of 1 volt will be seen for $V_0 = 10$ V. (With a capacitor of this size, the reading on the voltmeter will typically change fairly slowly.)

You could also plot out the speed and acceleration of the projectile from when you let go, allowing for acceleration by the air pressure and deceleration by friction in the tube and by the covering paper at the end of the tube. You

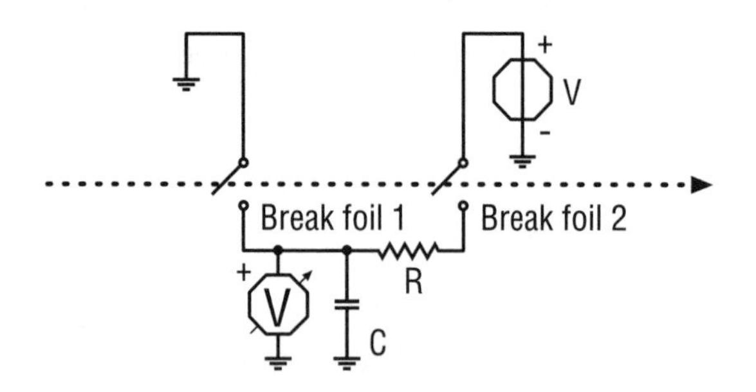

could put this along with a suitable model for how the tube pressure varies from the maximum partial vacuum to a minimum as the projectile compresses the air remaining in the tube. (In general, the remaining air is not sucked away fast enough by the vacuum cleaner.)

Finally, armed with the knowledge of the acceleration down the tube, you could redesign the launcher, perhaps employing a combination of "vacuum reservoir" (simply a large volume connected to the barrel to reduce the loss of vacuum as the projectile progresses up the tube), angled hinged flaps, and other features. With a means of measuring muzzle velocity, you could then check how much improvement these make.

REFERENCES

Holt, L.T.C. *Isambard Kingdom Brunel.* London: Penguin Books, 1957–1989 (various editions).

Jones, D.R.H. *Engineering Materials 3: Materials Failure Analysis.* Oxford: Pergamon Press, 1993.

Vaughan, Adrian. *Isambard Kingdom Brunel, Engineering Knight-Errant.* London: John Murray, 1991.

Sounds Peculiar

10 *String Radio*

> A murder has just been committed at Salt Hill and the suspected murderer was seen to take a first-class ticket for London by the train which left Slough at 7:42 pm. He is in the garb of a Quaker, with a brown coat on, which reaches nearly to his feet; he is in the last compartment of the second first-class carriage.
>
> —Telegraph message, 1845

The public was astonished when newspapers reported in 1845 the first known example of a criminal caught with the aid of a telegraphic message. The telegraphic message traveled at close to the speed of light (300,000 km/s) while the 7:42 train from Slough would never manage even 30 m/s (it was of course hauled by a steam engine at that date). In 1910, the world was even more astonished by an early feat of wireless telegraphy (radio). The notorious murderer Dr. Crippen and his accomplice, Miss Le Neve, were arrested on a transatlantic ship on the basis of a radio telegraph signal using a primitive Marconi transmitter.

In the String Radio demonstration, we input vibrations along the axis of a fairly taut string. The principal mode of sound propagation is longitudinal waves with a speed that is of the order of 1,500–3,000 m/s, in this case. This is remarkably fast compared to the speed of sound in air (about 330 m/s) and is of course much faster than the fastest possible train or airplane.

If electrical telegraphy and radio had not been discovered, it would still in principle be possible to catch the likes of Crippen or the Quaker, even if they were fleeing justice on the supersonic Concorde airliner, using sound waves in solids. Of course, there would be snags. For example, you would need amplifier devices every 100 km or so, even using monocrystalline quartz fibers to transmit the sound—and a transatlantic telephone call would involve a two-hour delay for the reply.

The transmission of sound through a solid is in general very complicated. The possibilities of sound transmission are both longitudinal (the solid is displaced along the line of transmission, like sound in air) and transverse (the solid is displaced at right angles to the direction of wave propagation, with up-and-down or side-to-side movement while traveling horizontally, called S-waves in seismic studies). Waves can be transmitted only on the surface of the solid (transverse again—with Love waves having motion parallel to the surface and Rayleigh waves having motion perpendicular to the surface, like water waves), or they can be transmitted through its bulk (like fast seismic or earthquake waves, P-waves). In thin sheets or strings, the speed of transverse sound waves is different again, depending on the tension in the surface or string: this is of course how drums and guitars (for example) are tuned. The speed of sound transmission can depend on as many as twenty-one independent elastic constants in an anisotropic solid such as a crystal or a piece of wood, giving different speeds in different directions, even for the same wave type. Fortunately, though, many solids are approximately isotropic, giving a single speed of sound for each mode of sound propagation.

Some solids also show a very low attenuation of longitudinal waves. The combination of their low attenuation with the fact that longitudinal waves are much faster in solids than they are in air is why the old Native American and Boy Scout trick of listening for trains coming along the steel railroad track is possible. Although placing your ear on the rail for this purpose is still not recommended, you will probably not be made into mincemeat by the engine arriving. For example, assume a source of sound of intensity of, say, 100 decibels (dB) at the engine, and a minimum audible level (with some distractions such as people telling you to get off the line) of 30 dB, and you have a 70 dB margin. Taking the value of attenuation in steel at the 500 Hz value shown in the table here, this would give 3.5 km as the maximum detecting distance. Even a high-speed train at, say, 200 km/h will take a minute to arrive at this speed, while the sound waves will arrive in just a second or two over the same 3.5 km. What is more, the sound in the railroad will go around curves in the track. Listening for the train via sound waves in the air, is, by comparison, a dumb idea: if a curve stops sound through the air until the train is 1,000 m away, for example, then you will have only eighteen seconds to react, and three of those valuable seconds will be taken up by the sound waves getting to you!

The table lists attenuation constants of some of the commoner plastics. The attenuation is measured in decibels per meter. For each dB of attenuation the sound will be 26 percent less in power than it was before, while for each 10 dB of attenuation the sound will be ten times less loud. The attenuation of longitudinal waves in many solids increases with frequency approximately linearly. I have calculated many of the figures in the table from measurements made at MHz frequencies, however, which may lead to some inaccuracy. Also, strings, even monofilaments, have been stretched, aligning the long polymer molecules from which they are made, and this lends them properties that are significantly different from those of the bulk isotropic cast plastic. For these reasons, the table must be taken as very approximate indeed.

The table lists the attenuation at both 500 Hz and 5 kHz, the latter being, on the approximation of linear increase with frequency, ten times that of the 500 Hz figure. With low frequencies attenuated less than high frequencies, there is a muffling effect—after a few meters, the music sounds as if the radio has been turned up to full volume but wrapped in blankets. The table lists the difference between the 500 Hz and 5 kHz attenuation as the muffling effect.

	Attenuation along 5 m path		
Material	*500 Hz (dB)*	*5 kHz (dB)*	*Muffling Effect (dB)*
Nylon	0.06	0.56	0.5
Rubber	0.93	9.31	8.4
Rubber/carbon	2.30	22.96	20.7
Neoprene	2.00	19.98	18.0
Perspex	0.50	4.95	4.5
Carbonate	1.04	10.42	9.4
Styrene	0.20	2.00	1.8
PVC	0.22	2.17	2.0
Aluminum	0.0009	0.0087	0.0
Steel	0.0107	0.1073	0.1
Quartz	2.76E-05	2.76E-04	0.0

Source: G.W.C. Kaye and T. H. Laby, *Tables of Physical and Chemical Constants*, 16th ed. (Harlow, UK: Longman, 1995).

The low attenuation values were well known to Victorian scientists and showmen. Professor Charles Wheatstone (inventor of the pre-Morse telegraph system named after him, as well as the Wheatstone Bridge, the potentiometer system for measuring resistance and a musical instrument, the concertina) carried out extensive investigations into sound waves. In 1840, Wheatstone coined the name "telephone" for a system for conducting musical notes a few tens of feet by wooden rods. A number of showmen staged special effects such as the "piano in the box." In this set-up, music from a genuine full-size piano in a secret room was conducted by a thin rod glued to the sounding board through a hole in the wall to a membrane in a small box in the audience. When the piano was played, the box appeared to contain a grand piano! In the days before electronics this type of demonstration made a big impression. Apparatus to do this sort of thing made by Professor John Tyndall around 1850 is still kept for its historical value at the Royal Institution in London.

The Degree of Difficulty

This project is not particularly difficult, although you need a little confidence to take the outside case off a transistor radio and solder some external wires onto the wires that normally connect to the radio's internal loudspeaker (although on many radios you can use a jack plug in the earphone socket instead). You will need to devote some effort to setting up the transmitting string so that it is taut, and to muffling the loudspeaker so that transmission of sound through the air is minimized.

What You Need

- ❑ Transistor radio
- ❑ Jack plug (to fit the radio earphone jack socket, if fitted)
- ❑ Monofilament nylon line (fishing line)
- ❑ Small loudspeaker
- ❑ 2 pieces of lightweight connecting wire
- ❑ Aluminum sheet, 0.5 mm
- ❑ Plastic disposable cups (the thin-walled kind that burns your fingers when it has hot coffee inside, not the Styrofoam kind)
- ❑ Metal or wooden stakes (just rods, pointed at one end, about 150 mm [6″] long)

- ❑ Solder and soldering iron
- ❑ Superglue
- ❑ Bricks, books, etc.

What You Do

Take the case off the radio. With a soldering iron, unsolder the two wires going to the loudspeaker and extend them with new pieces of wire you take outside the case. Then you can put the radio back together, but don't screw the case back so tightly that you cut the two wires. Alternatively, solder two wires to a jack plug of the correct size to fit the radio's earphone socket.

Solder the two wires onto your loudspeaker. A loudspeaker magnet and frame are the fixed parts that need to be anchored down, while the moving coil moves the cone in and out, and this needs to pull on the line. Now glue one end of a 5m piece of the nylon line onto the edge of the central cone of the loudspeaker (other positions in the middle will also work).

Attach the loudspeaker in a vertical position behind a hook or stake stuck in the ground (or fixed to a base if you are doing this indoors). To minimize the direct sound from the loudspeaker through the air, surround it with walls of books or bricks, covering them with a blanket or similar material. You could

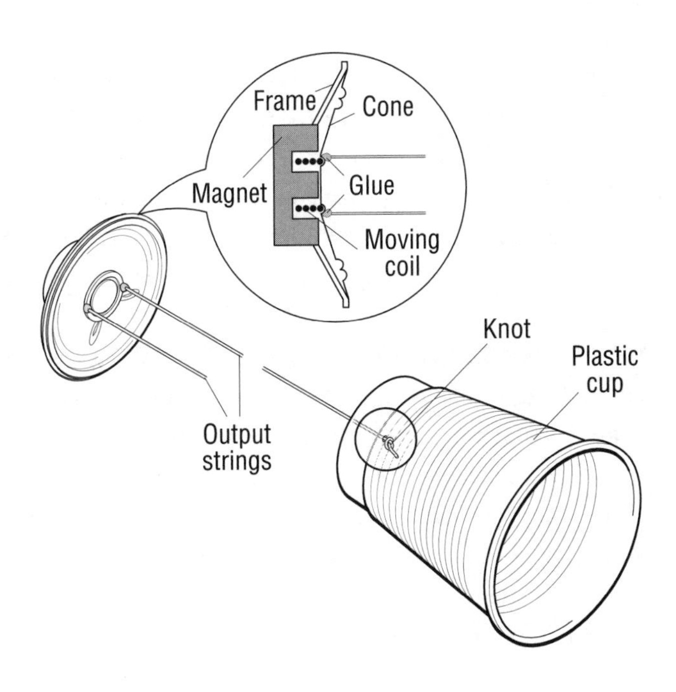

even glue circles of firm plastic to the front and back of the loudspeaker to reduce the amount of sound energy going to the air, putting a hole in the front to let the nylon line out. Make sure the line is not touching anything and exits freely straight from the loudspeaker.

Make a small hole (just big enough for the string to go through) in the middle of the coffee cup base. Thread the free end of the string through the hole and tie a couple of knots so that the cup cannot come off.

Now tune the radio to a suitable channel and adjust the volume so that only faint sounds can be heard coming from the loudspeaker. Position the coffee cup at the side of your head so that you can gently tighten the string. Music should emerge!

The Tricky Parts

You may need to support the string so that it does not droop to the ground. You can do this by tying or looping a piece of support string underneath it, at right angles, and suspending the transmitting string from it—but don't forget that the transmitting string must be taut.

The Surprising Parts

The tension must be carefully adjusted to give reasonable volume and quality without wrenching the loudspeaker cone from its mountings. With no tension, you can hear nothing. With surprisingly little tension, the signal grows to a reasonable volume. Further increases in tension seem to increase not volume but quality (perhaps high frequencies get through better; there is less rattling). The support strings might be expected to absorb a lot of the sound—but in fact they don't seem to.

I used a 0.7 mm monofilament polymer, a kind of fishing line, for my tests (breaking strain of 20 kg force), and, although it purported to be made of nylon, it certainly attenuated much more than would be indicated by the figures shown in the table from G.W.C. Kaye and T. H. Laby. With a 5 m path length, I could detect a significant muffling effect—certainly more than 3 dB (a factor of 2). On the plot given, I have shown the muffling effect on a typical audio spectrum (0–5 kHz), with 20 dB attenuation at 5 kHz, showing the downward shift of the peak, which our ears interpret as muffling.

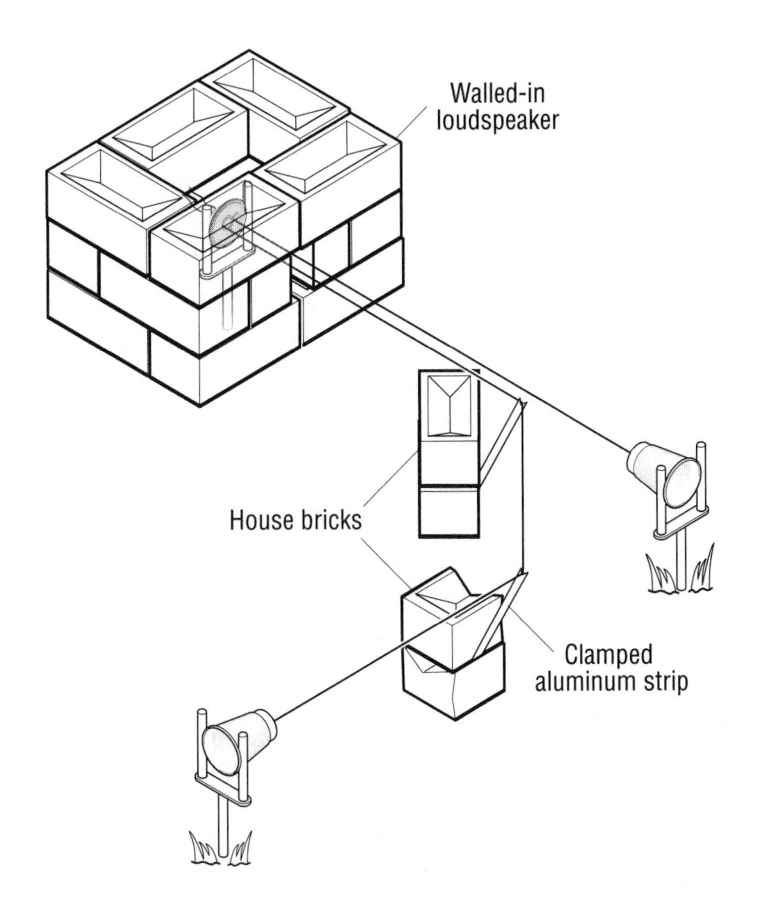

Using Your String Radio

One of the surprising things is that it is possible to take the string around corners without losing the signal. I used small strips of 0.5 mm thick aluminum, cut to about 10 mm wide with an obtuse V-angle in the tip, and aimed to bisect the angle through which the string is to be turned. I found that a turn through 45 degrees worked perfectly well, and that two turns in series did not produce a problem either.

Naturally, you could use the pull of a knotted string to provide a turn—but at some risk that the signal will be shared with the turning string. This is certainly the case if the turning string is at a small angle to the outgoing string—this can clearly be used for splitting signals.

What is the maximum turn angle that can be accommodated in this approach?

The Short Explanation

The corner-turning ability of the String Radio signal is explained if the motion of the aluminum corner-turning piece is considered. When a longitudinal wave reaches the metal, it pushes it back and forth, and it then in turn pushes the next piece of string back and forth. The only losses in the process occur because

- if the metal is heavy, then it will not move much, and the wave will be reflected. With thin aluminum, this is not the case, and the wave is largely transmitted.
- the aluminum is flexing in a direction that makes an angle of about $A/2$ (where A is the angle of deflection of the string) with the incoming and outgoing strings. There will be a small component of the string's longitudinal wave that is longitudinal to the aluminum plate, and a small part of the acoustic energy will be propagated along the plate, probably to be lost by absorption in the support.

THE SCIENCE AND THE MATH

A wave traveling along the string has an equation such as

$$Y = A \exp[2\pi fi(t - X/c)],$$

where Y is the displacement of the string (longitudinally) from its rest position at any point along the string distance X from the origin, A the amplitude of the wave, f the frequency, i the square root of -1, t the time, and c the wave speed.

This describes a sinusoidal wave that progresses along at speed c.

If the string has a discontinuity, such as a change of string material at the aluminum turning piece or where the string divides into two, the wave will not be transmitted at 100 percent amplitude but will split between the strings at the node, some of it being reflected. For a simple example, suppose the string has a joint in it, with a new piece of different speed c': there will be one wave in (above) and two waves out (transmitted and reflected waves, amplitude T and R) from the joint. Now unless the string is going to pull apart at the joint, the displacement of the joint due to the incoming wave must be equal to that of the outgoing waves. So

$$A \exp[2\pi fi(t - X/c)] + \mathrm{R} \exp[2\pi fi(t + X/c)]$$
$$= \mathrm{T} \exp[2\pi fi(t - X/c')].$$

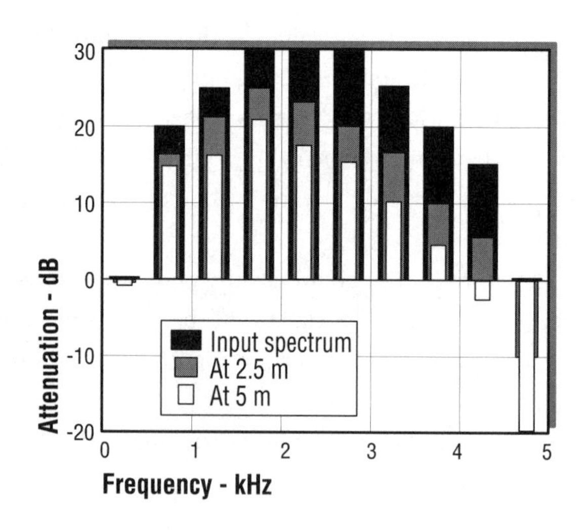

Now there is another condition that must apply across the joint: the joint's speed must be the same on both sides of the joint. This means the differentials with respect to time of the waves must equal each other. Therefore,

$$2A\pi(f/c)\exp[2\pi fi(t - X/c)] - 2R\pi(f/c)$$
$$\exp[2\pi fi(t + X/c)] = 2T\pi(f/c')\exp[2\pi fi(t - X/c')].$$

These equations can only be satisfied for all time if

$$A + R = T$$

and

$$(A/c) - (R/c) = T/c'.$$

What this means is that if c is nearly equal to c', then $T = A$ and $R = 0$ (good transmission, no reflection); if c' is much larger or smaller than c, then $T = 0$ and $R = T$ (poor transmission, most energy reflected). So joining a heavy string onto a lighter string, or vice versa, is likely to cause much of the signal to be lost.

The absorption of the sound wave along the string can be allowed for by adding a small real part to the imaginary part in the exponential: the wave amplitude falls off exponentially.

$$Y = A\exp(aX)\cdot\exp[2\pi fi(t - X/c)]$$

or

$$Y = A\exp[aX + 2\pi fi(t - X/c)],$$

where a is the absorption coefficient giving the length in which the sound reduces in amplitude by a factor of e, the base of natural logarithms.

And Finally, for Advanced Users

Several string radio outlets can be taken from one small loudspeaker cone, gluing them to the front of the moving coil, spaced evenly around the circle (the illustration earlier shows two lines attached). A little more cunning in the arrangements would no doubt yield stereo sound! The volume of sound from the string system (which is limited by the rattling of the coffee cup as you turn up the radio volume) can be improved by using a larger membrane at the receiving end. Children's toys similar to tennis rackets but with a clear plastic membrane instead of strings work well. One brand, designed to make a noise when used, is called BoomBats.

You could also use something less sophisticated than a plastic coffee cup as the output device: your teeth! Sound is conducted through your teeth and bones—jaw and skull—fairly efficiently, so you can hear music through your teeth by biting on the end of the string.

Distinct echoes are difficult to produce on the string: absorption is relatively high, and the speed of sound in the string is sufficiently high that an unfeasibly long piece would be needed for echoes to be clearly heard. However, the system does appear to have quite a bit of reverberation on it. Some of the reverberation and buzzing sound derive from the coffee cup earphone. However, you will hear reverberation from the string—probably caused by multiple reflections. Try adding a Y-branch, for example, near the walled-in loudspeaker, and you will hear the reverberation arising from signals that go and come back up the Y.

It is also possible to make the String Radio work with transverse waves. The loudspeaker is arranged to pull sideways near one end of a stretched string, while a light piece of balsa wood a few millimeters long and 1 or 2 mm on each side is glued sideways onto the string near the opposite end. The coffee cup earphone is, in turn, glued onto that. I found that at reasonable values of tension this worked quite well, although not as well as using longitudinal waves. The transmitted sound was more distorted—presumably by guitar string resonances at lower frequencies. Also, it is more difficult to support the string without losing sound, although horizontal support strings with vertical waves worked well (allowing the system to transmit around corners). One problem that is not seen with longitudinal waves is that the speed of sound in the transverse case varies with the tension in the string. As tension varies, "wowing" of the transmitted signal (frequency shifts due to changes in resonant frequency and momentary slight shifting of frequencies) occurs.

REFERENCES

Appleyard, Rollo. "Charles Wheatstone," chapter 4 in *Pioneers of Electrical Communication.* London: Macmillan, 1930.

Braddick, H.J.J. *Vibrations and Waves.* New York: McGraw-Hill, 1965.

11 *Mole Radio*

Ohm has taught us regarding the laws of the current. . . . The quantity of current is inversely proportional to the resistance. A clear image of the process is derived from the deportment of water. When a river meets an island it divides, passing right and left of the obstacle, and afterwards reuniting. If the two branch beds be equal in depth, width and inclination, the water will divide itself equally between them. If they are unequal, the larger quantity of water will flow through the more open course. And, as in the case of the water we may have an indefinite number of islands, producing an indefinite subdivision of the trunk stream, so in the case of electricity we may have, instead of two branches, any number of branches, the current dividing itself among them, in accordance with the law which fixes the relation of flow to resistance.

—John Tyndall, discourse at the Royal Institution of London, June 17, 1879

John Tyndall described the laws of current as succinctly as anyone could today, and all this before electricity had escaped from the laboratory. Ohm's famous law was first published in 1827. The law applies not just to current in circuits but also to electric currents that are fed into a continuous medium such as the Earth. This can readily be appreciated by picturing an indefinite subdivision of circuits. To stretch Tyndall's analogy further, we can imagine the positive electrode as a waterfall falling onto a flat plain, the water current splashing out sideways and then flowing away, with the negative electrode a sink hole sucking down water from the shallow water a little distance away from the waterfall. And if DC currents can flow, then AC currents can flow, and messages, and even, as in the Mole Radio project, currents modulated by speech and music.

In days of yore (before transistor radios but slightly after Robin Hood and his Merry Men), it was essential to have elaborate antennas and ground wires for a radio set. A massive 1 m copper rod was hammered laboriously into the ground, and then 100 m of wire was festooned around a bunch of nearby trees. J. A. Fleming, in *Electric Wave Telegraphy and Telephony* (pages 569–570), describes some of these effects, and how ground conductivity was studied in the 1880s as a possible communication system.

The result was, of course, a louder signal for the ordinary radio station listener. This arrangement, if not tuned to a station, also tended to receive weird radio signals from the atmosphere, called static or VLF (Very Low Frequency). These signals, caused by distant lightning, sound like the chirruping and whistling of peculiar birds, although all this was accompanied by a loud buzzing or humming deriving from the domestic AC electric supply.

However, this wasn't all: users of radio sets would often discover they could also hear the (highly distorted) telephone conversations of their neighbors. In this case, the signals came not from radio waves at all but from direct conduction, not along wires as a telephone works, but via conduction of electricity through the ground. This is the effect we demonstrate here.

Underground life forms on planet Earth are simple creatures. But if intelligent moloids had developed technology comparable to us humanoids, perhaps the Mole Radio is how they would have transmitted their news and pop songs. Try it out in your yard to see how well it works. And can you receive more than one channel? In our yard, moles with a radio set have a choice of two subterranean channels.

With the aid of two radio sets or tape recorders, some wires, and an audio amplifier, you can have your very own Mole Radio system in your yard—and no radio license is required.

What You Need

- ❑ Transistor radio set or tape recorder with earphone socket (or two, for two channels)
- ❑ Thin single-conductor insulated connecting wire (20 m per channel for transmitters, 2 m for the receiver)
- ❑ Metal stakes (I used 200 × 10 mm diameter [8″ × 3/8″] steel for the fixed stakes, and 100 mm [4″] iron nails for the movable ones.)
- ❑ Alligator-clip connectors
- ❑ Audio amplifier and loudspeaker*

What You Do

Arrange the transmitter as follows: hammer two large metal stakes into the ground as in the diagram, running the wires unobtrusively around the edge of

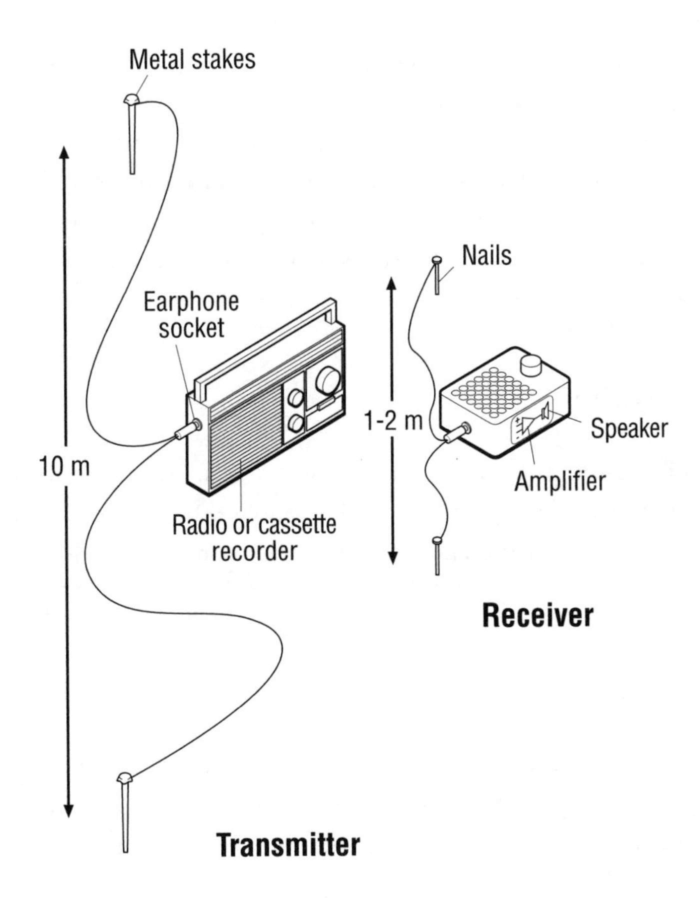

*These are sometimes available as a specific module, for example, as a guitar practice amplifier. However, a converted radio set or an intercom or a simple integrated circuit amplifier with battery and speaker from Radio Shack or a similar store would do.

the yard, if possible, and clipping them onto the stakes with alligator clips. The transmitter's (tape recorder or radio) earphone socket is used to push a powerful audio frequency current into the ground. If there is no earphone socket, then open up the case, disconnect the wires from the loudspeaker, and extend them (see previous chapter).

Wire the audio amplifier receiver to two metal rods (large nails seem to work quite well). The length of wire used is a matter of convenience, but I found that two wires about 2 m long gave quite enough signal using a simple single silicon chip audio amplifier. Push the rods into the ground in a line parallel to the desired station's electrodes and connect the wires using the alligator clips. The loudest signal is obtained by placing the receiver rods the greatest distance apart.

The Tricky Parts

Variations in the ground conductivity mean that some adjustment may be necessary to avoid getting a mixture of both channels. You have to orient the receiver electrode axis to maximize the signal.

The efficiency with which the arrangement works may decrease in dry weather. U.S. East Coast moles will probably get much better radio reception than will Arizona moles. In drier conditions, perhaps it will help to wet the ground around the transmitter electrodes with a bucket of water or two (as recommended for old-fashioned army field telephones).

The Surprising Parts

That the Mole Radio works at all may be a surprise. In fact, the sound is remarkably clear, although a little power line or household main current hum (60/50 Hz and its harmonics from the main AC electricity supply) does break through, typically.

Using the Mole Radio

You can move around the yard, simply pushing in the small rods to get a local connection wherever you like. You do not even need a volume control—just widening the spacing between the electrodes increases the volume.

It may help to know the best way to orient the receiver nails. You can consult the table in this chapter, which shows electric potential in volts near two electrodes at ±1000 mV under ideal conditions. The table is plotted on a graph to give a better feel for its behavior.

The signal receiver is maximized if the nails are placed along a steep slope, so that the voltage difference is greatest. The table lists the actual potentials in mV on the 20 × 20 m area. It is clearly seen on both that if the nails are oriented along the axis joining the two transmitter electrodes, or on the curves shown approximately on the graph, then little signal will be picked up: it has, in radio parlance, been "nulled out." With that transmitter nulled out, another transmitter channel with differently oriented electrodes will be heard well. Clearly the converse enables the first channel to be heard if the second is nulled out.

Another fact is highlighted by the numerical plot: there are substantial signals to be had well away from the two transmitter electrodes. If one says that

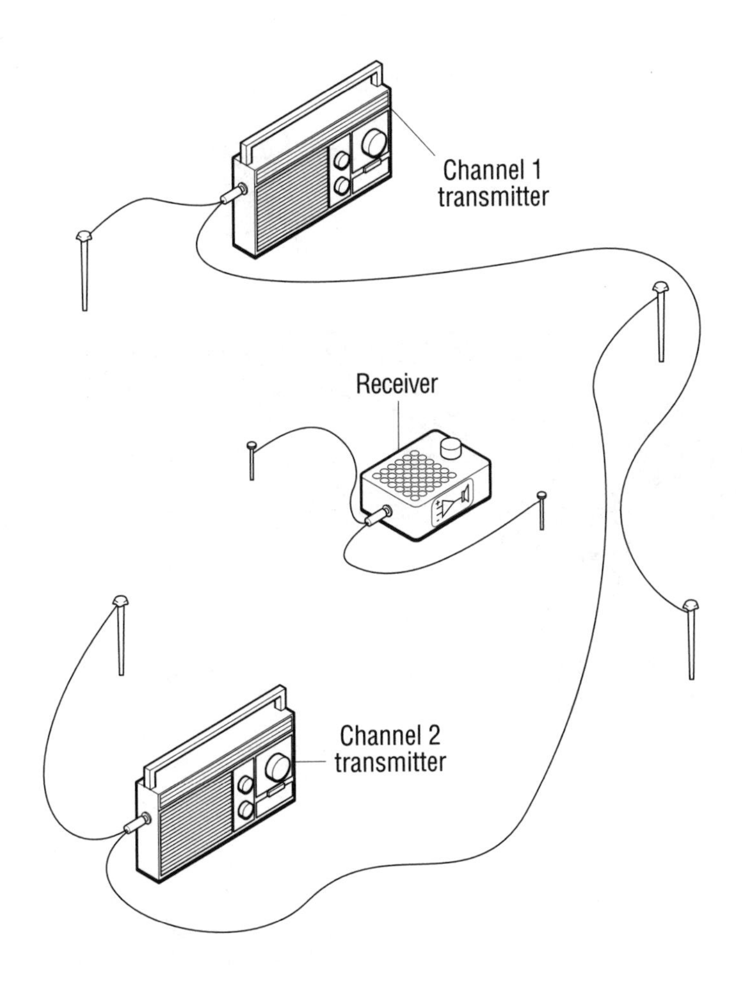

a 10 mV difference is required for a clear signal, then it is clearly quite possible to receive the transmitted signal at the corners of this plot, even with receiver electrodes still only 1 m apart. (The plot, incidentally, is a diagram discovered by Apollonius in the third century, sometimes known as "Apollonius's circles.")

The mathematics of the plot are very simple: the electric potential difference V due to two points with voltage difference $2V_0$ is simply given by remembering that the electric potential near an electric charge Q is kQ/R, where k is a constant.

The electric potential at any point due to the two charges is then $k(Q_1/R_1 + Q_2/R_2)$. Since the charges must be equal and opposite, and adjusting the constant k suitably, we have

$$V = V_0/(1/R_1 + 1/R_2),$$

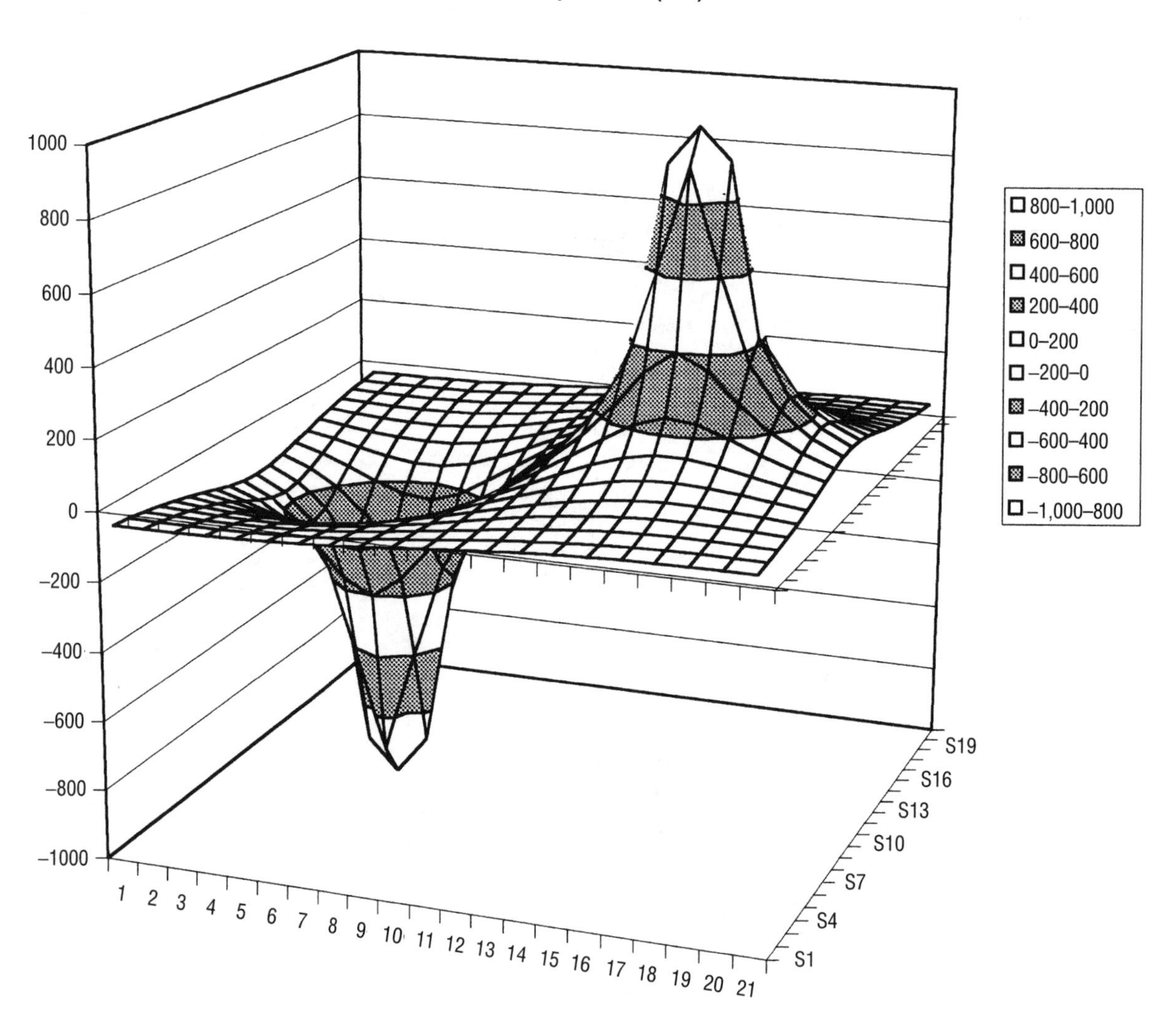

where R_1 and R_2 are the distances from electrodes 1 and 2, given in this case by a Pythagorean formula, that is,

$$R_1 = \sqrt{(X - X_1)^2 + (Y - Y_1)^2 + \ldots} \; .$$

These are the potential differences that are seen in the absence of any electrical conduction in the ground. However, if the ground is homogeneously conductive, then these are exactly what will be seen. This is because the electric charge density is almost instantly equalized in a conductor. Local electric fields due to nonuniform charge are soon eliminated by conduction (within a millionth of a picosecond in a metallic conductor, and not much slower in moist soil). Only with conductivity on the order of picomhos (picosiemens) per meter (high-quality insulators) can charges build up within a solid. With a uniform charge density, and a net charge of zero, the flow of current through the resistance of the solid determines the voltage differences seen.

The generalization of Ohm's law is

$$J = \sigma E,$$

where J is the current density vector, σ the conductivity, and E the electric field.

Furthermore, the definition of electric potential V is that

$$E = -\text{grad } V.$$

Grad V is the three-dimensional gradient of V, so that

$$\text{grad } V \equiv i\partial V/\partial x + j\partial V/\partial y + k\partial V/\partial z.$$

Now if the current out of every little element of the soil is net zero (as much current goes in as goes out), then div $J \equiv \partial J_x/\partial x + \partial J_y/\partial y + \partial J_z/\partial z = 0$.

Now in an isotropic medium, σ is a constant everywhere, if div $J = 0$, so div $E = 0$, so

$$\text{div } (-\text{grad } V) = \nabla^2 V = 0,$$

$$\text{where } \nabla^2 V \equiv \partial^2 V/\partial x^2 + \partial^2 V/\partial y^2 + \partial^2 V/\partial z^2.$$

This is exactly the same equation (Laplace's equation) as for electrostatic potential in free space.

Variations in the ground conductivity will distort the patterns of electric potential seen on the surface of the ground considerably, since if s is not constant in the equations just given, div $J = 0$ does not imply div $E = 0$, and Laplace's equation will not apply. To take a trivial example, if the patch of soil between the two receiver electrodes is much more highly conductive than the surrounding soil, then the potential differences between electrodes in the outer area will be tiny.

And Finally, for Advanced Radio Moles

Variation in ground conductivity is used by geophysicists for prospecting for mineral deposits, as described, for example, by D. S. Parasnis in *Principles of Applied Geophysics* (pages 101–130). Water, conductive ore bodies, and even, indirectly, oil, can be detected hundreds of meters down with favorable conditions. It is also what archaeologists use when they carry out resistivity surveys of ground when prospecting. Anthony Clark, who wrote *Seeing beneath the Soil,* described to me how he developed early "Avometer" (multimeter) methods into a usable prospecting method that he and then thousands of other archaeologists have gone on to use extensively, finding and mapping buried ancient sites across the world.

Electric Potential in mV from Two Electrodes at ±1,000 mV Placed 10 m Apart

–34	–35	–35	–34	–32	–29	–25	–20	–14	–7	0	7	14	20	25	29	32	34	35	35	34
–40	–41	–42	–42	–40	–37	–32	–25	–18	–9	0	9	18	25	32	37	40	42	42	41	40
–47	–50	–52	–52	–51	–47	–41	–33	–23	–12	0	12	23	33	41	47	51	52	52	50	47
–56	–60	–64	–65	–65	–61	–54	–43	–30	–16	0	16	30	43	54	61	65	65	64	60	56
–66	–73	–79	–84	–85	–81	–72	–58	–41	–21	0	21	41	58	72	81	85	84	79	73	66
–78	–89	–100	–109	–113	–111	–99	–80	–55	–28	0	28	55	80	99	111	113	109	100	89	78
–92	–108	–126	–145	–157	–157	–141	–112	–76	–38	0	38	76	112	141	157	157	145	126	108	92
–106	–130	–161	–197	–229	–238	–211	–160	–104	–51	0	51	104	160	211	238	229	197	161	130	106
–120	–153	–201	–271	–358	–402	–339	–232	–140	–65	0	65	140	232	339	402	358	271	201	153	120
–130	–171	–240	–364	–617	–900	–597	–323	–175	–78	0	78	175	323	597	900	617	364	240	171	130
–133	–179	–256	–417	–909	–V	–889	–375	–190	–83	0	83	190	375	889	+V	909	417	256	179	133
–130	–171	–240	–364	–617	–900	–597	–323	–175	–78	0	78	175	323	597	900	617	364	240	171	130
–120	–153	–201	–271	–358	–402	–339	–232	–140	–65	0	65	140	232	339	402	358	271	201	153	120
–106	–130	–161	–197	–229	–238	–211	–160	–104	–51	0	51	104	160	211	238	229	197	161	130	106
–92	–108	–126	–145	–157	–157	–141	–112	–76	–38	0	38	76	112	141	157	157	145	126	108	92
–78	–89	–100	–109	–113	–111	–99	–80	–55	–28	0	28	55	80	99	111	113	109	100	89	78
–66	–73	–79	–84	–85	–81	–72	–58	–41	–21	0	21	41	58	72	81	85	84	79	73	66
–56	–60	–64	–65	–65	–61	–54	–43	–30	–16	0	16	30	43	54	61	65	65	64	60	56
–47	–50	–52	–52	–51	–47	–41	–33	–23	–12	0	12	23	33	41	47	51	52	52	50	47
–40	–41	–42	–42	–40	–37	–32	–25	–18	–9	0	9	18	25	32	37	40	42	42	41	40
–34	–35	–35	–34	–32	–29	–25	–20	–14	–7	0	7	14	20	25	29	32	34	35	35	34

Note: Connecting up points of equal potential gives a contour diagram of the electric potential. Each of these contours would represent a horizontal slice through the three-dimensional diagram shown earlier.

They survey a site typically by pushing a pair of fixed reference electrodes into the ground, then move around with a pair of sample electrodes, measuring the potential difference between a fixed electrode and a moving electrode, while injecting current into a fixed electrode and a moving electrode. The moving electrodes are inserted at a set of grid positions.

Archaeologists see both positive and negative anomalies. Filled-in ditches, for example, are usually higher in conductivity (they have a higher water content than the surrounding, more compact soil), while stone walls are lower in conductivity than the surrounding moisture-saturated soil. Resistivity anomalies may reverse, however, depending on weather conditions.

Could measurement of resistivity using transmitter electrodes at right angles as in our demonstration and, say, three or four receiver probes provide additional information for the archaeologist?

Finally, over what distance could messages be sent with the system? Arthur C. Clarke describes in *Voices across the Sea* the incredible struggle against the odds to lay a cable underneath the Atlantic: eight years of toil were needed to complete a successful link. Could pioneers like Lord Kelvin and Cyrus Field have saved themselves a lot of trouble by taking a cable from London to Cape Wrath (northwest Scotland) and planting electrodes? A similar installation would take a cable from the Florida Keys to the Hudson Straits via New York. But would it work? And how much power would be needed? And what would be the effect on the Atlantic Ocean? Sea water, after all, is quite a good conductor, and, worse still, there are ocean currents of the liquid flow kind as well as the electric kind. Seawater flowing in the magnetic field of the Earth gives rise to DC electric potentials, which have even on occasion been used to measure flow speed.

REFERENCES

Clark, Anthony. *Seeing beneath the Soil: Prospecting Methods in Archaeology.* 2d ed. London: Batsford, 1996.

Clarke, Arthur C. *Voices across the Sea: The Story of Deep Sea Cable-Laying.* London: Frederick Muller, 1958.

Fleming, J. A. *The Principles of Electric Wave Telegraphy and Telephony.* London: Longmans, Green, 1916.

Parasnis, D. S. *Principles of Applied Geophysics.* 3d ed. London: Chapman and Hall, 1979.

12 *Bat Doppler*

Twinkle twinkle little bat,
How I wonder what you're at!
Up above the world you fly!
Like a tea-tray in the sky.

—Lewis Carroll,
Alice in Wonderland

Bats do not twinkle, as Lewis Carroll suggests, in the sense of emitting pulses of light energy like distant stars in the night sky. However, they do "twinkle" sound waves of 20–80 kHz, above the limit of human hearing. They emit pulses of ultrasound and use the echoes as a kind of navigational and hunting radar—more correctly, sonar. Bats use the pulse echo timing and the echo phase to estimate the distance and direction of objects in their path. Bats also tune their emitted pulses to avoid interference with neighboring bats (see Glenn Zorpette, "Chasing the Ghost Bats"). Whether bats use the Doppler shift in the echoes is not mentioned in the books on bats I found. (Perhaps any reader who can shed light, or even ultrasound, on this will let me know.)

The curious fact that frequencies of sound waves do not remain a constant when emitted from a source that moves was first described by Christian Johann Doppler in 1842. He and his associates carried out experiments (these sound like a lot of fun) involving people on railroad carriages with trumpets. Surprisingly soon after Doppler's findings were published, Frenchman M. Hippolyte Fizeau proposed that light frequencies would also show Doppler effects. This required a big leap of imagination: the sound waves of Doppler are measured in wavelengths of centimeters and frequencies of thousands of cycles per second, whereas light waves are measured in fractions of a millionth of a meter and have frequencies around 10^{15} (1,000,000,000,000,000) cycles per second. Surprisingly, perhaps, Fizeau was shown to be correct only twenty years later when the astronomer Sir William Huggins discovered the Doppler effect in light ("redshift") in the spectrum of the star Sirius. Curiously, the Doppler effect in ultrasound was discovered later in history, simply because ultrasound was not fully investigated until late in the nineteenth century. R. Koenig showed that if you made a tuning fork shorter and shorter it still worked as a tuning fork and still emitted sound. But the sound got too high to be heard as the prongs on the fork became very small: what we now call ultrasound. But this is not a good way to generate ultrasound and the paper in the references was not published in 1899.

The use of an ultrasonic carrier frequency allows the Doppler effect to be heard much more clearly than it can be at ordinary audio frequencies. This is because the absolute amount of the Doppler shift is proportional to the frequency used. For example, if a sound source moving at a walking pace, 1 m/s, emits a frequency of 1,000 Hz, then the Doppler shift will move the frequency heard by observers in front of the source by only 3 Hz, to 1,003 Hz, a musical interval of only one-twentieth of a semitone. If the same sound source were emitting 40,000 Hz, the Doppler shift would be 120 Hz. The shift of 120 Hz would be readily detectable as two musical semitones were it presented as a shift of the 1,000 Hz tone.

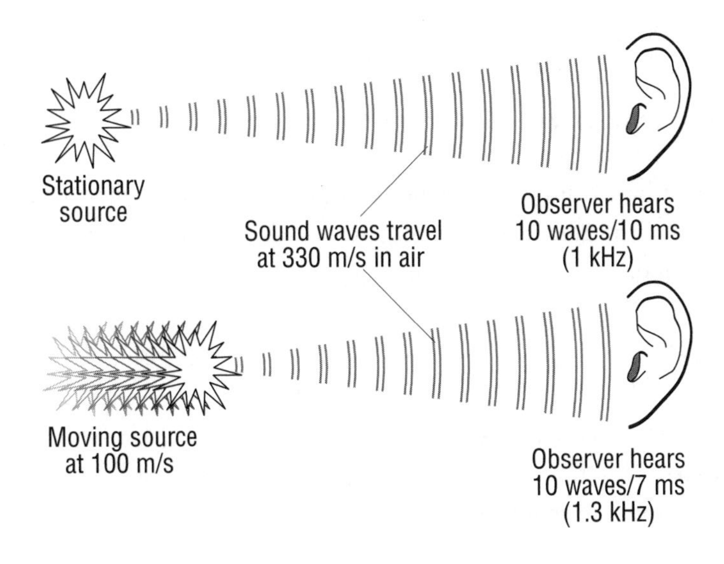

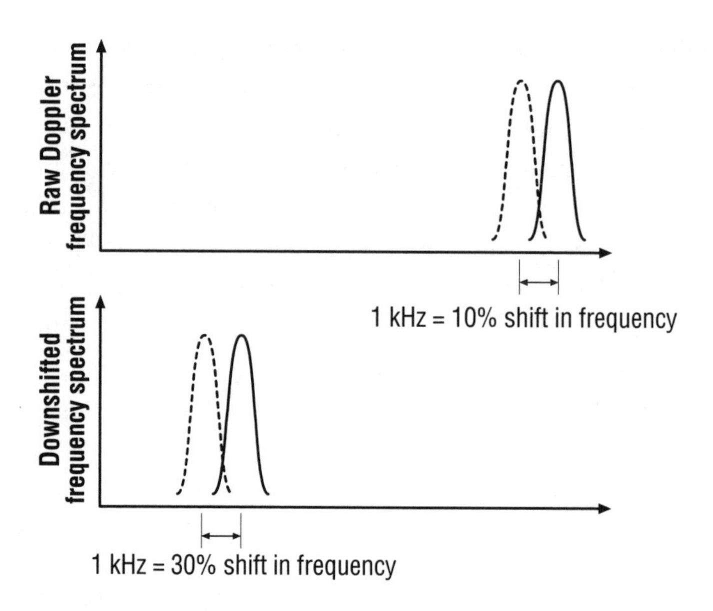

To avoid the need for listeners to possess bats' ears, and to impose the Doppler shift produced on a lower frequency to increase the relative shift of frequency, a frequency-shifter circuit is used in the Bat Doppler. This sort of circuit could have other uses, of course. By tuning input frequencies up a little, rather than down as we do in this demonstration, parents could raise the whining tone of small children to a supersonic pitch at which it would be inaudible. Conversely, of course, by tuning down somewhat less than is necessary for bat detection, children could get their own back by sending messages secretly using ultrasonic dog whistles!

The Degree of Difficulty

Any electronics buff will breeze through this project. The untransistorized, however, may need some guidance about circuit boards, soldering, and so on, to get this project going.

What You Need

- ❑ 3140 op amp integrated circuit*
- ❑ 2 integrated circuits, 555 oscillator
- ❑ 2 batteries and connectors, 9V PP3

*The best bet for the electronic gear is a mail-order house like Newark Electronics (Farnell Electronics in the UK), or an electronic hobbyist store like Radio Shack.

Bat converter

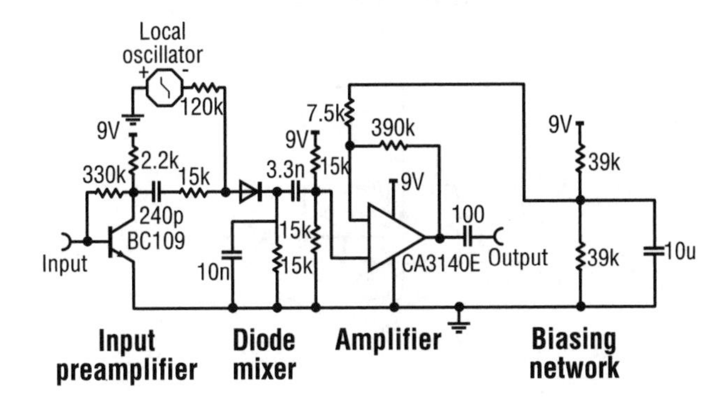

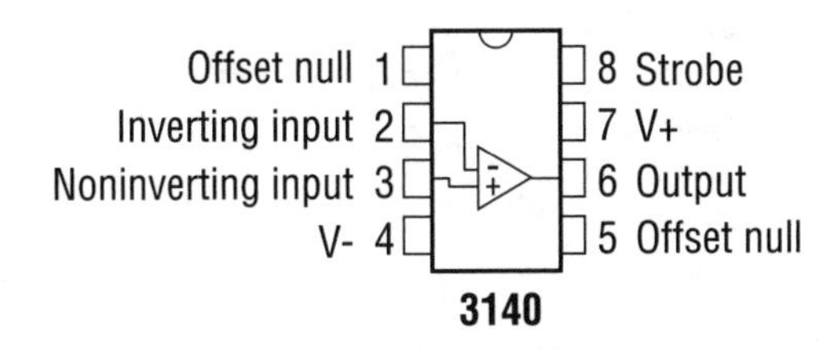

- ❑ Ultrasonic transmitter and receiver pair (I got mine for free by cannibalizing a burglar alarm.)
- ❑ Audio amplifier and speaker
- ❑ Transistor BC109
- ❑ Diode (1N914 or anything similar)
- ❑ Resistors and capacitors (see circuit diagram)

What You Do

The circuit comprises two parts, the Cyberbat and the Bat Converter (BC). The Cyberbat emits a steady ultrasonic note at 40 kHz (which you cannot hear). This note will be shifted up or down by the Doppler effect. The Bat Converter converts the ultrasound to regular sound at around 1,000 Hz (which you can hear), but with the same Doppler shift as the ultrasound. Note that the Bat Converter includes an electronic oscillator identical to the oscillator of the Cyberbat.

Switch everything on and place the Cyberbat a few feet away from the BC. You should hear (in general) a high whistling sound. Whatever kind of tone you hear, tune the BC oscillator until the sound is a comfortable middling kind of

Oscillator for cyberbat ultrasound emitter and local oscillator for bat converter

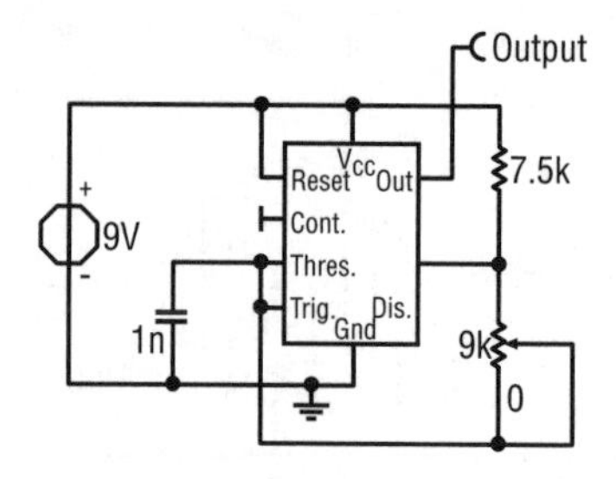

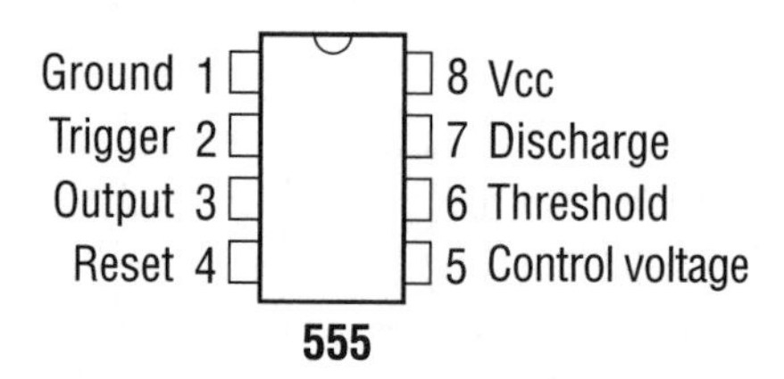

tone. Now try moving the Cyberbat (or the BC) toward or away: you will find that only modest movements serve to give increases (motion toward) or decreases (motion away) of musical pitch.

The Tricky Parts

You do need to have all the connections on the circuit board present and correct—and no extra connections. (If you are stuck with a nonfunctional circuit, an oscilloscope of course is a boon, particularly as the two oscillators here are ultrasonic.) It is also an advantage to tune the oscillator of the Cyberbat so that you get the strongest possible output from the ultrasonic transducer, which will only be achieved over a narrow band of frequencies, perhaps with 1 kHz of the 40 kHz resonant frequency of the transducer. However, when you retune the Cyberbat, you will also need to retune the BC to match.

The Surprising Parts

I think you will find this works much better than ordinary demonstrations of Doppler: small movements of the Cyberbat are instantly audible, even for the tone deaf, as large changes in musical pitch. Also important, the changes

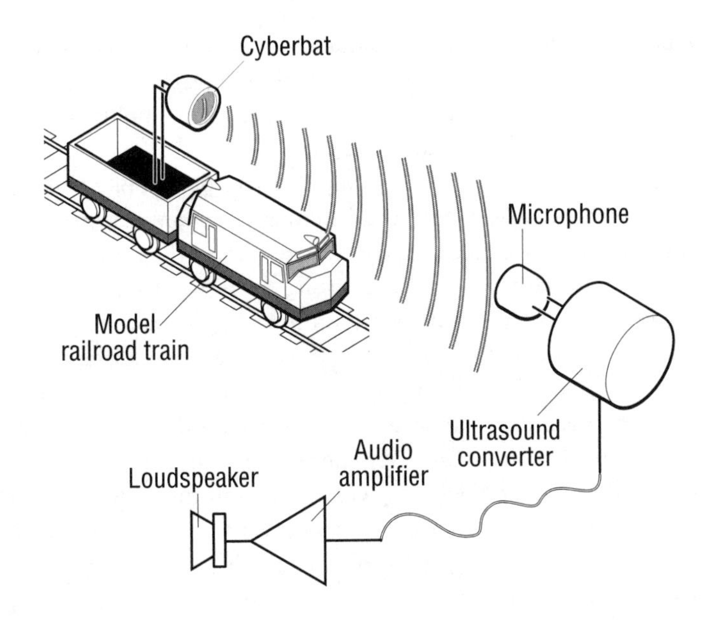

in loudness that confuse the listener in ordinary Doppler experiments are minimized.

Using the Double Doppler

The musically minded might like to try this out on their model train set. By setting up a long straight section of track and turning up the volume on the amplifier only when the train with the Cyberbat aboard is on the straight heading directly for the Bat Converter, you will hear a fairly well-defined tone. Remember that each musical semitone amounts to a change of frequency by a ratio of the twelfth root of 2 (1.05946, or about 6 percent). If your ear is not good at hearing absolute tone, try listening to the Cyberbat briefly with the train stationary, then turn down the volume while the train gets up to full speed and turn the volume up again. In this way you can clearly hear the tonal interval, a much easier musical task.

THE SCIENCE AND THE MATH

The Cyberbat is an oscillator with a transducer to convert its output into sound waves, which are a steady ultrasonic note at 40 kHz. The Bat Converter converts the ultrasound to regular sound at around 1,000 Hz. It contains a microphone to convert the received ultrasound to electricity, and after amplification it adds the ultrasonic microphone signal with a similar but constant frequency signal at 41 kHz (the "local oscillator" signal). When you add two frequencies that are similar, you get a difference frequency output that can be separated by a diode/capacitor combination such as the BC includes. Suppose two signal waveforms Y and Y are added, at frequencies F and G, to form a new waveform Z: the waveform Z is a high-frequency $(F + G)$ wave that is modulated in amplitude from 0 to 1 by a low-frequency modulation wave at frequency $F - G$. The diode combination eliminates all the negative excursions of the high-frequency wave, producing a row of peaks from zero to a height that varies slowly. The capacitor averages out the peaks to leave a simple wave at the difference or beat frequency $F - G$. Mathematically, you can simulate the diode approximately by assuming the diode gives an output that is the square of the input:

$$X = X_0 \sin (2\pi F)$$

$$Y = X_0 \sin (2\pi G)$$

$$Z = X + Y = 2X_0 \sin (\pi(F + G)) \cos (\pi(F - G))$$

$$Z' = Z^2 = 4X_0^2 \sin^2 (\pi(F + G)) \cos^2 (\pi(F - G))$$

But

$$2 \sin^2 A = 1 - \cos (2A),$$

so

$$Z^2 = X_0^2 [1 - \cos (2\pi(F+G))] [1 - \cos (2\pi(F - G))].$$

But the average over a short time of the high frequency $(1 - \cos (2\pi(F + G))$ is simply 1, because the cos term averages to zero. The second term is then left, and it is a simple sine wave at frequency $F - G$ with a 90-degree phase shift (hence cos) and an offset upward, so that it goes from zero to 2 in value:

$$Z^2 = X_0^2 [1 - \cos (2\pi(F - G))].$$

The beat or difference frequency at about 1 kHz (plus or minus the Doppler shift of the Cyberbat), once separated out by a diode (the mixer) and filter, is now at audible (audio) frequencies; it has been down converted. It is then amplified for output to a loudspeaker.

The Doppler effect or Doppler shift is very well described in almost every physics textbook. But here is a quick explanation. If the transmitter of a sound wave is emitting f waves per second normally, each will travel at the speed of sound C. The wave crests will thus be spaced apart by a distance of C/f. But if the transmitter is made to move toward the listener at speed V, the wave crests will be spaced apart by $C/f - V/f$, since in the intervals of time $1/f$ between emitting crests, the transmitter will have moved distance V/f. These wave crests continue to travel at speed C toward the listener, who hears wave crests spaced apart by $(C/f - V/f)$ traveling at speed C or, in other words, a sound of frequency $C/(C/f - V/f)$, that is, a higher pitch of sound f', where

$$f' = f/(1 - V/C).$$

In this standard case, therefore, the key quantity is how great the speed is compared to the speed of sound, which is quite fast, about 300 m/s in normal air. For example, taking a middle C tone of 440 Hz, to get a frequency shift of one whole musical tone, or 12 percent, you need to have a speed of 36 m/s, which is quite a high speed, the speed of an automobile on a fast highway. By using the ultrasonic frequency of 40 kHz, and then subtracting 39.56 kHz of that, we need a speed of only 0.4 m/s to get a whole musical tone of the Doppler shift.

Automobile burglar alarms (the kind that go off only after the burglar has broken in) often use the Doppler effect with ultrasound. One transducer emits a steady 40 kHz tone, and another receives the transmitted wave after reflection off the inside parts of the automobile, and, of course, that reflected off the burglar. An attenuated version of the transmitted tone is added to the receiver output tone, and the added waveform is rectified and somewhat smoothed at a time scale of a few tens of Hertz.

This new waveform includes of course the beat frequency of the transmitted tone and the reflected and double Doppler shifted tone from anything that is moving inside the automobile. Another circuit detects whenever this beat frequency tone occurs and sounds the alarm siren.

And Finally, for Advanced Users

You can probably most easily detect the doubled Doppler effect, which happens when sound reflects off a moving target, by constructing a corner reflector. The inside corner of a rectangular box has the curious property that its three surfaces mutually at right angles will reflect any incoming ray (after one, two, or three reflections) back along an exactly parallel path. The corner reflector is most efficient when the incoming light enters along what might be dubbed the double diagonal or three-dimensional diagonal line—the line that goes between opposite corners of a cube, as opposed to the ordinary diagonal between the opposite corners of a square.

I made such a reflector simply by soldering together three triangles of copper-faced fiberglass-reinforced circuit board. If you make the reflector accurately and 15 or 30 cm on each side, then at up to 5 or 10 m you should be able to pick up the Doppler shift due to moving it. The transmitter and detector need to be side by side, pointing in the same direction, but perhaps with a screen between them. Corner reflectors can be almost 100 percent efficient in reflecting power back to the emitter illuminating them, and this is why they are used for radar reflectors at sea, so that small vessels can be seen more clearly on radar than the tops of waves, even in stormy weather.

You could try making two Cyberbats and listening to the strange harmonies that result as you wave them around in the air, or as they move on a model railroad. The sound can be reminiscent of a pair of theremins. These objects, invented by a Russian, Leon Theremin, were the world's first electronic musical instruments, built back in the 1920s using thermionic tubes. From time to time since their invention they have been used to create weird "science fiction soundtrack" music. Could these Cyberbats be competition for the theremin?

Finally, by using an ultrasonic microphone with a wider bandwidth (the one suggested has a very narrow band, giving the benefit of a strong rejection of interference), you will be able to hear a range of ultrasonic sources in the environment. You will be able to hear, for example, the deafening ultrasonic whistle made by TVs and computer screens, and by dog whistles and burglar alarms. Depending on where you live, you can also hear bats calling and emitting their echo-location clicks. You don't have to go to Belize in the Caribbean, as Glenn Zorpette did, either. Much less exotic locations, like the edge of most towns on a summer night or in parks near trees, are often worth a try. Several electronic-gadget kit suppliers sell gadgets for bat detecting.* There are also professional bat detectors designed for connection to your portable computer—the Anabat II is an example of these (see Zorpette, "Chasing the Ghost Bats"). These kits will of course also work with our Cyberbat to demonstrate the Doppler effect.

REFERENCES

Bergman, Ludwig. *Ultrasonics and Their Scientific and Technical Applications.* Translated and expanded by H. Stafford Hatfield. New York: Wiley, 1943.

Koenig, R. "Ueber die hoechsten hoerbaren und unhoerbaren Toene von 4096 bis 90,000 Schwingungen" (On the highest hearable and unhearable sounds from 4096 Hz to 90 kHz). *Ann. Phys. Lpz.* (III) (1891): 69, 626, and 721.

Zorpette, Glenn. "Chasing the Ghost Bats." *Scientific American,* June 1999, 74–81.

*An Australian company, Titley, makes the Anabat professional bat detectors: *http://www.titley.com.au.* A simple bat detector is available from Magenta Electronics Ltd., 135 Hunter Street, Burton-on-Trent DE14 2ST, UK. Tel. + 44 (0)1283 565435; *http://www.magenta2000.co.uk.*

Transmissions with Omissions

13 *Toothless Gearwheels*

> The improved mechanical efficiency of machines still depends upon a further improvement in gear manufacture.
>
> —Sir Joseph Whitworth, 1862

Gearing is often needed for making gadgets.* Unfortunately, making gears and transmissions is very tricky. Have you ever thought about making a gearwheel? You need to cut an exact circle, then notch it with precisely shaped notches all the way around—with the notches matching up when you get back to the beginning again. And you have to make all of this in a hard-wearing and thus hard-to-cut material. Then, to use it, you need to make another gearwheel whose teeth precisely engage the first, despite being a different diameter, mount both on shafts, fix the shafts so that they rotate but don't slide in the transmission, and finally, if all is well, connect your input power and output shaft. It is all but impossible, except with heavyweight dedication or some fairly sophisticated manufacturing machinery.

It is surprising to learn, given that gearing is so difficult, that gears are thought to be very ancient. Gearwheels are believed to have been first used in ancient China, perhaps as early as 200 B.C., for use, for example, in harnessing the power of oxen to irrigate crops. Earlier gears are thought to have been used by the Greeks in instruments such as the antikythera, an astrolabe device for measuring and predicting star and planetary positions. The Roman Empire made

*Sir Joseph Whitworth, quoted in the chapter epigraph, was a pioneer of precision engineering and standardization and promoted standard screw threads, some of which are still named after him. Sir Robert Hooke and Camus had made similar statements about the need for gear improvement more than a century earlier.

use of gears too, both in power applications, such as milling, and in control applications, such as distance measurement—the odometer. A design for an odometer carriage is contained in the works of Vitruvius in the early first century B.C., incorporating multiple gears and a reservoir of pebbles that drop through a hole in the final gear every mile. Metal working in the Roman era was still relatively primitive, with hand-beaten wrought iron expensive, difficult to work, and inconsistent, while easily worked alloys like bronze lacked strength and wear resistance. This ingenious machine was long thought impossible for Roman metallurgy to have managed, but Andre Sleeswyck in "Vitruvius' Odometer" reported reconstructing it successfully.

Gearwheels became common in Europe with the growth in use of waterwheels and windmills in the medieval and Renaissance periods, but mostly using an unsuitable combination of wood and metal. It must be said in the defense of the use of these materials, however, that the best metal (brass) and the best wood (lignum vitae) are a very good combination for low-power mechanisms such as clocks. John "Longitude" Harrison (see *Longitude* by Dava Sobel) was making his first astonishingly accurate chronometers for the British Royal Navy using wooden parts in the 1720s. The wood was self-lubricating and did not corrode in the presence of salt spray.

Gearwheels suffered from problems in materials, manufacturing, and physics even as late as the era of Queen Victoria, however. Gears need to be made of materials that are easy to machine as well as wear resistant and strong, superficially opposing requirements in those days. Only in the last 150 years or so has the technology developed that allows gears to be cut from soft annealed steel alloys that can then be hardened by heat treatment.

Lubricants were needed to reduce wear and frictional losses, but at first only crude animal and vegetable products were available. The accuracy of manufacture of gears—the constancy of tooth pitch and the regularity of tooth form—was also a big problem. Whitworth was typical of engineering pioneers in machinery, and he faced problems typical of them—not only the usual shortages of money, the usual choruses of Doubting Thomases and NIH-ers (not invented here), but also the "pulling yourself up by your own bootstraps" problem. Without accurate parts, you can't make accurate machines to make accurate parts, and so on.

Finally, the physics of the action of gear teeth—how they partly roll on each other, partly slide, how the different mathematical shapes (involute, cycloidal) work, how rotational speed can be modulated in gearing—were all imperfectly known. These factors were major restraints on nineteenth-century engineering, and solutions to the problems were delayed, perhaps surprisingly, until well into the twentieth century.

Gearing is most often used to increase the torque, the amount of twisting action available from a power source. Electric motors, as we have mentioned earlier in this book, typically produce rather low torque value. This can be increased (at a cost of slowing the output-shaft rotation speed) by gearing down. Conversely, there are sources of reasonably high torque, but these need to be speeded up or geared up. An example might be the mechanism of a Bourdon pressure gauge or a barometer: a small change in air pressure gives a small movement in the sensor element—seldom more than a couple of millimeters—which needs amplifying up to 50 mm or so to make it readily visible.

Gearwheels are normally made with teeth, which have to slide on each other, at least to some extent. The sliding action is undesirable in that it wastes energy, it can wear the gears out too soon, or it may require the use of oil lubrication, with all the problems that entails.*

But the gearwheels don't need to touch each other! Instead of tangible, solid teeth, gearwheels can use invisible "magnetic" teeth. The acme of the gearwheel designers—transmission without wasteful sliding action—is attainable, and in this demonstration we will find out how.

*There are a number of types of gear or gearwheel. A *contrate* (crown-shaped) *gearwheel* has teeth pointing parallel to the shaft, while a *spur gearwheel* has teeth pointing radially outward. A *pinion* is a small spur gearwheel. *Bevel gears* have teeth pointing at 45 degrees to the shaft, and a pair of them allows power to be transmitted around a 90-degree bend. *Helical gearwheels* are spur gearwheels with teeth aligned along a helical track (giving more efficient and quieter transmission, especially in large machines). A *worm* is essentially a screw designed to engage with one or several teeth of a spur gearwheel, so that one turn of the worm moves the spur gear one tooth along. However, every one of this cornucopia of mechanical gears involves some sliding.

The Degree of Difficulty

Making Toothless Gearwheels is straightforward, although obtaining an ideal ring magnet is relatively difficult. In a broken HP IIP laser printer, in the mirror-rotator mechanism, I found a ring magnet with north and south poles arranged around its perimeter, with a steel plate stuck on the back to enhance its magnetic flux on one side, as the diagram shows. This kind of ring magnet is easier to find than you might think. Dismantle an old floppy-disk drive or maybe even a computer hard drive and you will find similar magnets. You also need a Meccano or Erector Set or a similar construction kit to be able to make axles that rotate reasonably smoothly and accurately while under some force.

What You Need

- ❑ 1 or more ring magnets (see the diagram) or a number of small identical bar magnets
- ❑ Meccano or Erector Set or similar wheels to mount magnets on

'Contrate' gearing 6:1 ratio

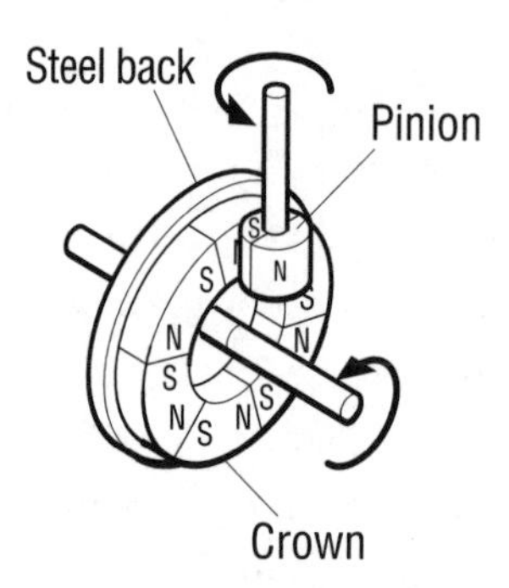

Alternate soft-iron gearwheel, 11:2 ratio

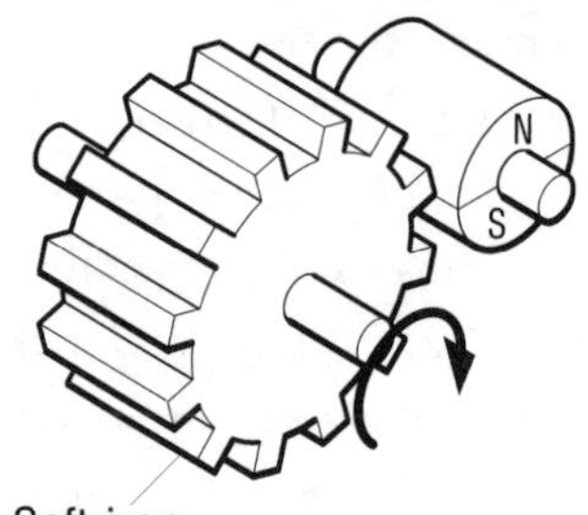

'Double contrate' gearing 5:3 ratio

Toothless bevel gears

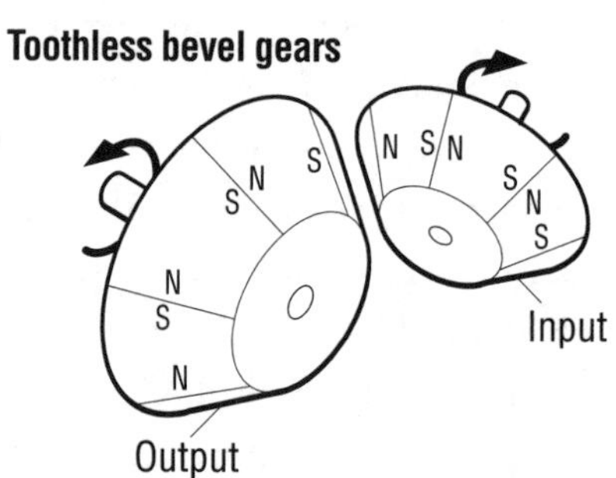

'Spur' gearing 5:4 ratio

Toothless worm gears

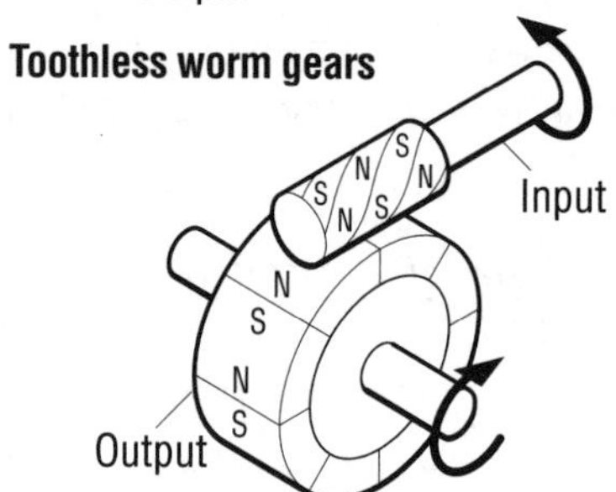

- ❏ Sundry Meccano or Erector Set or similar parts
- ❏ Input crank or input motor with battery
- ❏ Set of identical permanent magnets (best are powerful neodymium-iron-boron or samarium-cobalt magnets)

What You Do

I made a contrate transmission, meshing the ring gear with a small rotating bar magnet. You can of course make such a ring magnet easily by gluing a set of small bar magnets (made perhaps by scoring and breaking a brittle rod magnet) onto a circular steel backing sheet. A similar set of spur gearing is simple enough to make with the same technique.

I arranged input power to come either from an electric motor with a bar magnet glued on the end, or from a simple hand crank.

The Tricky Parts

Of course there are problems. The magnetic teeth cannot resist much force, limiting the maximum torque available from the arrangement. The use of powerful (but more expensive) samarium-cobalt or neodymium-iron-boron magnets in the smaller gear helps here, though. Also, in a multistage reduction transmission, the first stages do not need very high torque, while the low friction characteristic of the magnetic cogs is a distinct advantage. The springiness of magnetic teeth helps too, in that configurations such as the double-contrate gearing in the diagram would tend to jam if the teeth were solid metal.

The gears are subject to surprisingly high forces, so you need to have a strong mounting frame: tape and cardboard are definitely ruled out, although I found that even a rather crude Meccano or Erector Set frame served admirably.

The Surprising Parts

You also need to restrain the meshing magnetic gears from meshing mechanically or flying apart, as I have explained. But it works! You can even space the toothless gears apart as much as ¼″ (although the torque produced will be small). And a magnetic tooth drive needs only two teeth on the driver pinion.

With a motor-driven gear set, the magnetic teeth work well at high speed—with low friction. Try slowing down the output shaft by gripping it with your fingers: the input shaft slows down too, which is not quite what you might intuitively expect.

The force between the two gear "teeth" will be, if you have used reasonably powerful magnets, about 1 Newton: the force exerted by 100 g (4 oz) in the Earth's gravity field. A handy way to visualize a force of 1 Newton is to think about the weight of an apple.*

*The apocryphal story about the discovery of the science of gravity is that Isaac Newton lay underneath an apple tree in his garden—it is still there, in a garden at Woolsthorpe, not far from Cambridge, England. He noticed an apple fall and figured out his theory, based on all bodies in the entire universe attracting all other bodies according to their mass and distance. (It seems singularly appropriate that the scientific unit of force named after him has a value about equal to the weight of an apple in the Earth's gravitational field.)

THE SCIENCE AND THE MATH

The magnetic toothless gears avoid all possible sliding friction. Ordinary gears (along with many other mechanical devices) are subject to sliding friction: as the gear teeth mesh, they slide somewhat. This is most obvious with a worm gear engaging a spur gearwheel. However, it is also the case with a simple pinion engaging a spur gearwheel, albeit to a much smaller extent. Ordinary gears are especially designed to minimize that sliding, but it is impossible to entirely eliminate it. "Pure rolling motion occurs only when the contact point between gear teeth occurs at the pitch point . . . every other point of contact results in sliding of one tooth on another" (Michel, Sadler, and Wilson, *Kinematics and Dynamics of Machinery*). The pitch point is the point where the pitch circle intersects the gear tooth; the pitch circles are imaginary circles that would, acting as rollers, give the same motion as the gearwheels.

Sliding contact is undesirable in moving machinery for a number of reasons. First, it wastes energy because of the frictional forces between the sliding surfaces. Second, as if this were not bad enough, the wasted energy is itself a problem because it manifests as unwanted heating of the rubbing parts. Third, the parts in contact will be subject to wear.

Of course, it is possible to make transmissions with rollers, toothless gears that simply press hard on each other to maintain contact, using the force of friction to their advantage. (These are used, for example, with rubber wheels, in audio magnetic-tape drives.) However, there are rather low limits to the torque that can be transmitted unless a high contact force is maintained, and this requirement puts severe restrictions on their use.

Another kind of transmission without sliding that can be provided by rollers is by incorporating them as the "teeth" on gearwheels or chains. Rollers make substantially friction-free gears in mechanisms such as chain roller drives and roller teeth gears. At low speeds, these have no sliding involved in the basic action. The rollers simply turn instead of the teeth sliding. (We ignore the small amount of sliding in the roller pivots.) At high speeds, though, the requirement for the roller to speed and slow down rapidly in rotational speed as the teeth engage it means that sliding will take place.

Rollers are, of course, not completely friction free either. The roller elastically (that is, not permanently) dents the surface on which it rolls, very slightly, so it continuously has to surmount a tiny hill of its own making just in front of itself, which takes energy. Not all of this energy is returned to the roller by the push of the dent straightening itself out behind the roller.

Magnetic teeth, in addition to offering no sliding friction, are superior in other ways to a straightforward friction roller drive:

- A roller drive does not position the output shaft positively, and the reduction ratio is not an exact ratio of whole numbers.
- The magnetic teeth give a small and alternating thrust on the gear shafts, whereas roller gearing has to exert a substantial and constant force on the gear shafts, resulting in more wear and energy loss in the gear shaft bearings.

Toothless gears sound really great! Their Achilles' heel is of course the limited power or torque they can exert. But what is the maximum power and torque that could reasonably be expected from a set of toothless gears? A simple way of estimating the force available between two magnets is to calculate the magnetic energy (E_m) stored in an imaginary magnetic spring between two magnets. The energy stored in a spring is given by the force F the spring exerts multiplied by the distance s over which the force acts, or, if the force varies, the integral $\int F\,ds$.

The magnetic spring energy will be comparable to the energy E_g stored in the magnetic field of the gap

$$E_m \sim B_m H_m \cdot V_m \sim E_g,$$

where V_m is the volume of the two magnets (if they are similar in size) and $B_m H_m$ the maximum BH product of the magnetic material. The BH product is a standard quantity quoted for magnets, indicating the amount of energy per unit volume stored in their magnetic field.

E_g can give the force F by dividing by a distance, such as the separation of the magnets L:

$$F \sim E_g/L \sim B_m H_m \cdot V_m/L.$$

With a cheap Alnico magnet, the BH product could be, say, 10 kJ/m^3. For teeth engaging at a distance of 1 mm with 5 cubic mm effective volume of magnet, this gives about 1 N force. With modern "rare-earth" magnets (samarium-cobalt or neodymium-iron-boron), the force between teeth can be improved markedly, by a factor of, say, five or ten times.

The driving toothless gearwheel is in effect a rotating- or traveling-wave magnetic field. This could be synthesized by arranging a set of electromagnet coils and phasing the switching of the currents in them suitably. But this would amount to a peculiar multipole permanent magnet electric motor! (You could even replace the driven toothless gear with a set of coils with currents flowing in them—which of course would make it into a peculiar kind of electromagnet motor, similar to an AC permanent magnet generator worked in reverse as a motor. Many small AC generators, like those on motorbikes, or the magneto ignition systems on chain saws, are permanent magnet devices.)

And Finally, for Advanced Users

Toothless gears are free from friction, but are they really perfectly efficient? What about air drag? What about magnetic hysteresis? Are these significant? What is the maximum torque you can achieve from your toothless transmission? What happens when you exceed the maximum torque? You may notice that when you approach the maximum torque the smoothness of the output drive becomes markedly worse.

As shown in the diagrams, you can try a number of variations on the toothless transmission, simulating many of the standard gear types. I found the contrate and—a type of gearwheel almost impossible with mechanical parts—a double-contrate gear worked well. But perhaps the spur magnetic gearing (which works well in toothed gears) would be superior.

It is not essential to use permanent magnets in both driven and driving gearwheels. One of the wheels could simply be a permeable iron wheel, perhaps machined with a typical cycloidal tooth form like a normal gear, but designed to mesh with the magnetic tooth gear.

To do this you need to put teeth back on toothless gears. You can drive a toothed steel gear with a toothless magnetic pinion, if the steel is a suitably "soft" magnetic material (transformer iron is ideal). The teeth of the iron wheel could of course be hidden in a plastic (casting resin, perhaps) encapsulation, if you wanted to maintain the toothless appearance. There are obviously economies to be made in the gearing by doing this, although the torque supplied will be lower (and I found this arrangement only seemed to work on motor-driven input if you gave the gearwheel a start—if you try it yourself you may see why this probably has to be so). There is a question, of course, of what the optimum tooth form is for such a gearwheel.

There is also the possibility of providing a back-up of ordinary metallic contact bearing if the cycloidal form of the ordinary gear is retained and both gearwheels are toothed: the gears would be noncontact only under small torque conditions, reverting to contact gearing if the torque becomes high. (Some precautions, perhaps plastic coating, would be needed to avoid problems from magnetic wear particles coating the permanent magnet gearwheel.)

The Ultimate Transmission

What is the ultimate noncontact transmission, with guarantees of zero wear? Well, there are some industrial machines in everyday use with noncontact magnetic bearings. Turbo-molecular vacuum pumps, for example, use magnetic levitation (maglev) bearings. A transmission could be made, perhaps for high-speed use, using magnetic bearings as well as magnetic teeth!

REFERENCES

Laithwaite, Eric R. *A History of Linear Electric Motors.* London: Macmillan, 1987.

Michel, Walter J., J. Peter Sadler, and Charles E. Wilson. *Kinematics and Dynamics of Machinery.* New York: HarperCollins, 1983.

Mollan, S. *Mechanism Design.* Cambridge, UK: Cambridge University Press, 1982.

Sleeswyck, Andre. "Vitruvius' Odometer." *Scientific American,* October 1981, 158–171. (The National Museum of Naples, Italy, has on display a working model of Vitruvius's odometer.)

Sobel, Dava. *Longitude.* London: Fourth Estate, 1996.

Steeds, W. *Mechanism and the Kinematics of Machines.* London: Longmans Green, 1940.

14 *Flying Pulleys*

Dangerous at both ends and uncomfortable in the middle.

—Ian Fleming

A high-powered version of the Flying Pulley I made using chainwheels thrashed around rather scarily when 200 W (⅓ horsepower) of electrical power was applied to the motor. The resultant transmission looked pretty dangerous in the middle and pretty uncomfortable at each end, which brought to mind James Bond author Ian Fleming's famous description of a horse!

The early makers of gasoline engines had to contend with the problem that almost all applications of their engines required a relatively low speed but a high torque. Similarly, one of the difficulties of making anything move electrically is that the most efficient converter of electricity to mechanical power is the electric motor, and the electric motor only works well at very high speed and with very low torque. But the manufacture of a transmission, as discussed earlier in this book, is an exacting business, requiring exact positioning of gearwheels to work. It is simpler to use pulleys or chainwheels, but here again intermediate shafts and their bearings are needed. What happens if you leave out those shafts and mountings? Well, in general, the reduction gear will not work. But there is one way: the Flying Pulley.

The Flying Pulley should prove very useful in making small toys and other simple powered devices and might conceivably be useful in larger gadgets that need to be built in the simplest way possible.

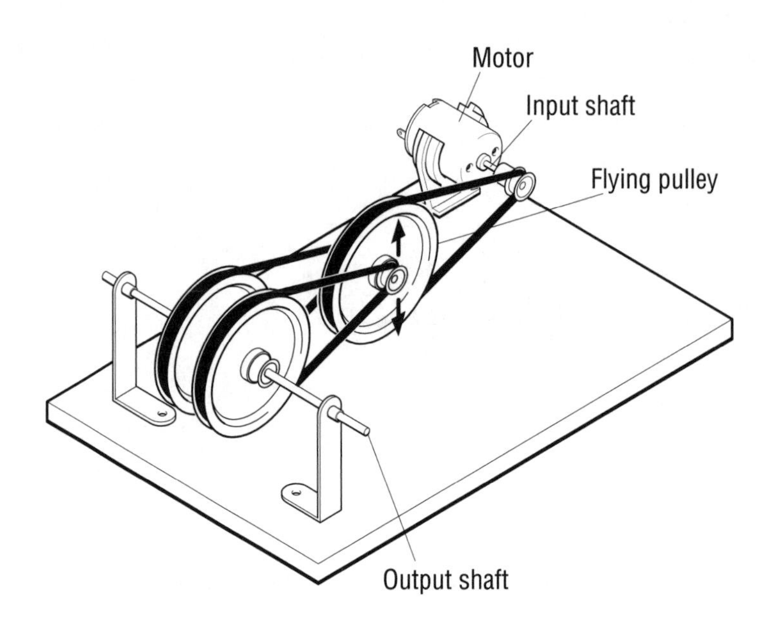

The Degree of Difficulty

This is a fairly easy demonstration, but you do need to choose your pulley bands, chains, or both carefully, and space the input and output shafts to match them.

What You Need

- ❑ Electric motor
- ❑ Batteries and battery box for motor
- ❑ 3 large pulleys or chainwheels
- ❑ 3 small pulleys or chainwheels
- ❑ Short stub intermediate flying shaft
- ❑ Output shaft on bearings
- ❑ Mounting to maintain relation of motor to output shaft
- ❑ Drive bands or chains
- ❑ Toothed belts and matching notched pulleys (for advanced version only)

What You Do

Assemble the parts as indicated in the diagram. Allow for some adjustment of chain tension by moving the motor and output shaft, and leave some room for the intermediate pulley or chainwheel cluster to move up or down. You may want to attach a "load" to the transmission. I used a large fan, but you could use a simple brake arrangement or make the transmission part of a model-vehicle chassis. Then adjust the chain tension and start up.

The Tricky Parts

If the tension in the pulleys or chains is insufficient, the assembly will tend to oscillate wildly when load is applied. As just noted, this can be a little scary when high power is applied. Even though I was using chainwheels, so the drive bands could not be thrown off, the system seemed intent on self-destruction unless some tension was applied.

Finally, the tension is taken by the motor bearings as well as by the output shaft, so it is important that the motor has reasonably good-quality bearings.

The Surprising Parts

How does the system actually apply an increased torque above what would be achieved by a single pulley? Where does the extra torque come from?

THE SCIENCE AND THE MATH

The flying pulley assembly works by applying a relatively large tension on the input pulley, and this is really its Achilles' heel. As a result of the increased tension (relative to fixed-shaft systems), there is more wear and more friction on the input-shaft bearings. The flying shaft deflects off the centerline by an amount that is just sufficient to ensure that the tension in the system will balance out the forces on the shaft.

The diagram shows how the torque and drive-band tensions might look in a system in which the drive bands are fairly elastic. You can picture the system as stationary.

$$\Delta T' = g\Delta T,$$

where $g = R/r$ = gearing ratio of idler pulleys.

The deflection angle α and the other quantities can be seen to be related approximately as

$$2T \sin \alpha \sim 2\Delta T'$$

thus $\sin \alpha \sim g\Delta T/T$.

How the system behaves depends on how elastic the drive bands are: if they extend easily under load, like ordinary rubber bands that are not too stretched, then the system will simply deflect sideways, none of the bands will run slack, and the math applies.

With rather inelastic drive bands, however, or with chains, the behavior will include the situation where the band or chain runs slack on the output side. In this case, the idler pulley will not deflect much from its static position when the system is

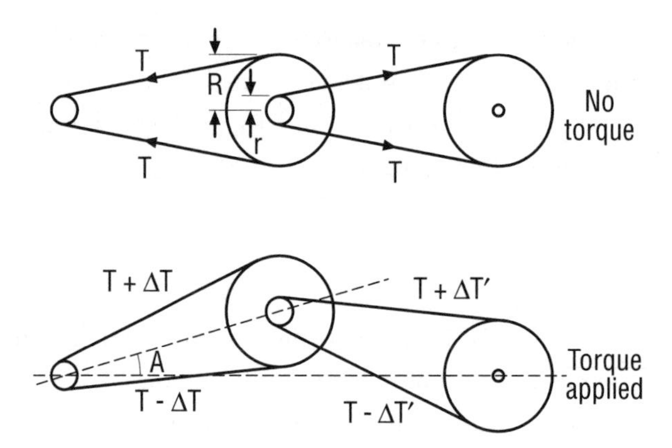

running, if there is little slack in the system. However, there will be a similar amount of tension on the input (motor) pulley bearings.

Engineers have long sought to eliminate bearings from machines, since in general bearing surfaces lead to wear on the bearing itself and on the axle that runs in it. There are other ways to avoid bearings that wear. Leonardo da Vinci once devised bearings based on rolling sectors (themselves based on earlier bearings based on rotating disks) for bell ringing, to make church bells that weighed up to several tons easier to swing, and to eliminate wear. Once metallurgy had developed sufficiently, rotating disk bearings led to the ball bearings and roller bearings that keep the wheels of industry turning today.

And Finally, for Advanced Users

If you have tried the simplest assembly with pulleys and simple elastic rubber drive bands, try the device using chains and chainwheels or, even better, toothed belts.

Flying Pulley transmissions may well have resonant frequencies at which they will thrash around and become inoperable. Interested readers are invited to try to identify and estimate the mass and spring components and then calculate these undesirable speeds of operation.

Clocks without Cuckoos or Quartz

15 *The Crank and the Pendulum*

> It had perceptibly descended towards me. I now observed—with what horror it is needless to say—that its nether extremity was formed of a crescent of glittering steel, about a foot in length from horn to horn; the horns upward, and the under edge as keen as a razor. Like a razor also, it seemed massy and heavy, tapering from the edge into a solid and broad structure above. It was appended to a weighty rod of brass, and the whole hissed as it swung through the air.
>
> —Edgar Allan Poe, "The Pit and the Pendulum"

This project is probably the world's simplest pendulum clock. There are no tricky escapements and gearwheel mechanisms to construct, making it very suitable for everyone, even torture chamber managers. The pendulum seems such a simple idea that it is astonishing to learn that there were water-driven, weight-driven, and even spring-driven clocks before there were pendulum-regulated clocks. Early mechanical clocks used much less effective regulating mechanisms, such as fan blades, rocking beams, and the like. Galileo observed in 1554 (before being arrested by the Inquisition for suggesting the Earth went around the sun) that suspended lamps in Pisa Cathedral swung with times that seemed to remain constant.

Galileo is said to have compared the lamp-swing time with his own heartbeat, which he thought of as steady. Of course the human heartbeat is not at all constant, and the variability of heartbeat is something we measure and use in medical diagnosis today. This again illustrates the sort of bootstrap problem faced by all pioneers. What gives the correct time, a pendulum clock or a human pulse? Later, Galileo compared pendulum swings with other phenomena—the motion of the sun and stars across the sky, for example, or the operation of sand hourglasses or water clocks. In this way Galileo (and his son, who built some of the first true pendulum clocks) eventually reached the conclusion that the pendulum clock gave a useful measurement of time.

The pendulum is, of course, only an oscillator. A true clock also has some means of counting the oscillations. To the simple pendulum, Galileo's son added mechanisms both for maintaining the oscillations (the "going train")—which would otherwise decrease to zero, thanks to friction—and for counting the oscillations (the "dial train").

The Degree of Difficulty

This is an easy project, if you make only the pendulum and its driving mechanism, and it is just slightly harder to make the clock operate a pair of hands. Unless you want to make your clock run to atomic accuracy or you want it to operate cuckoos and bells, you won't find anything impossible here.

What You Need

- ❑ Small 3–6 V DC electric motor (A cassette tape recorder type is ideal.)
- ❑ 10 ohm resistor (With different motors, a smaller resistor may be needed.)
- ❑ Battery box and battery (ideally rechargeable)
- ❑ Piece of wood for mounting (see diagram)
- ❑ Wheel or disk to fit the motor shaft or weight shaft, 20–30 mm diameter, drilled on the edge to make the driving crank
- ❑ Light wooden or balsa rod, 300–600 mm long
- ❑ Round nail, about 2″ long, 1.5 mm diameter, and washer or spacer that fits it
- ❑ Sundry tape and glue
- ❑ Light spring (such as 8 turns of 0.3 mm [30 AWG, 29 SWG] spring wire, or the key spring from a computer keyboard)
- ❑ A weight, about 50 g (such as modeling clay) for the pendulum bob

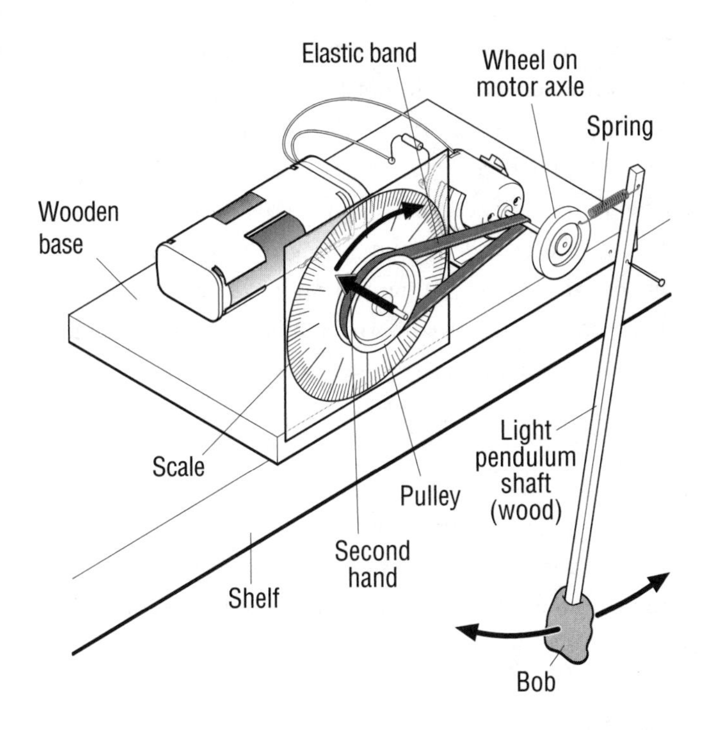

- ❑ Pulley, perhaps a plastic sewing-thread spool
- ❑ Nail or heavy-gauge wire to make an axle for the pulley
- ❑ Rubber band
- ❑ Cardboard scale/clock face and pointer
- ❑ Metal rod to make clock-hand shaft
- ❑ Metal bracket to support shaft

What You Do

Drill a hole in the wheel about 5 mm or less from the center for the crank. Attach the wheel to the motor and mount the motor on the wooden base. Attach the modeling-clay pendulum bob to the rod, make a hole at the other end of the rod for the spring, and then nail the rod onto the wooden base, using the spacing washer and ensuring that it swings freely. Connect the spring to the motor, connect the battery box via the resistor, and install a cell in the battery box. You now have the oscillator and going train of the clock complete.

You can also make a pendulum like this one to run on weights: instead of the electric motor, you will need a shaft on bearings. Wrap a piece of thin string with a weight on the end around the shaft (or "drum," if you have a larger diameter). The weight is wound up onto the shaft or drum by rotating the crank by hand. The shaft or drum should not be too large a diameter, and the weight should be modest, or the crank disk will overpower the restraining pendulum and the weight will plunge toward the floor with a frantic waggling of the pendulum. I used a 4 mm diameter shaft, which could run the clock only for about 1 minute, using a modest weight (about 25 g of modeling clay). Clearly, if you use a transmission that provides many turns of the crank for each turn of the drum or shaft, then you will need correspondingly more weight, but the clock will run for longer, for instance, 10 minutes with 10:1 gearing up and a 250 g weight.

The Tricky Parts

The spring must be connected to the pendulum correctly. If the spring biases the pendulum too far to the right or left, or if it is simply too stiff, the clock will not function correctly.

The Surprising Parts

With a bit of luck, unless the motor is a rather powerful one, you will find that when you give the pendulum a starting swing of the right amount, your clock will begin to run. The bob will be swinging back and forth regularly while the motor rotates sedately and irregularly. If the motor is too powerful, you may find the motor going around fast and furiously, while the pendulum stays vertical but wriggles frantically. The use of the 1.2 V rechargeable battery and the resistor should ensure that even powerful motors will not have enough torque to do this.

Pendulum Basics

A system like a pendulum has a part, such as the pendulum bob, that is being pulled toward a neutral position but can swing to either side of this position. It has a characteristic frequency, the resonant frequency, which is determined by two constants in the system:

- It increases with the strength of the restoring force, which is applied to the oscillating part as it moves from the neutral position.
- It decreases with the resistance to acceleration (inertia) of the oscillating part.

In the case of a pendulum, the inertia of the bob is simply its mass, while the restoring force constant is given by the force on the bob from gravity. This is simply its weight times the length of the pendulum times the sine of the angle that the bob makes with vertical. The weight is of course simply the mass of the bob times the acceleration due to gravity, which at the surface of the Earth is 9.81 m/s^2. The effect of the pendulum bob mass thus cancels out. A pendulum with a lightweight bob will swing at the same speed as a similar pendulum with a heavy bob.

In fact the resonant frequency of a simple pendulum (one in which a relatively heavy bob is fixed to a relatively light and long rod) is proportional to the square root of (the gravitational acceleration on the Earth divided by the length of the rod), for small angles of swing.

Without the pendulum, the motor would rotate either fast (if supplied with enough current) or not at all (if the friction in the motor was greater than the current supplied). However, with the pendulum, provided that more than a small minimum current is supplied, the motor will rotate, but only at one slow regulated speed. Unless it is very powerful (in which case, it will go around fast and furiously), the motor will run only at the regulated speed because the pendulum acts as a "resonator," which will allow oscillation to occur only at its resonant frequency.

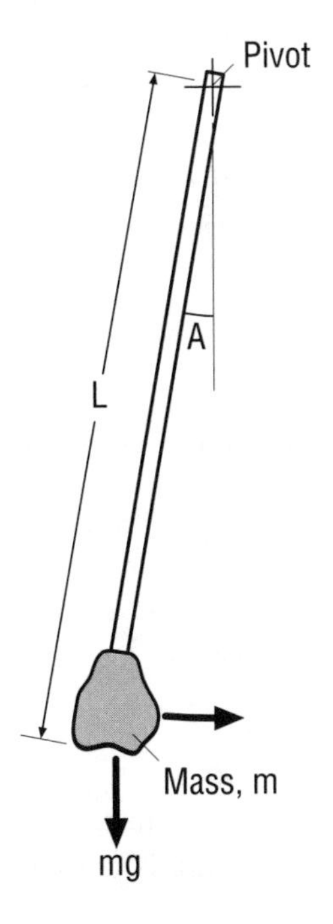

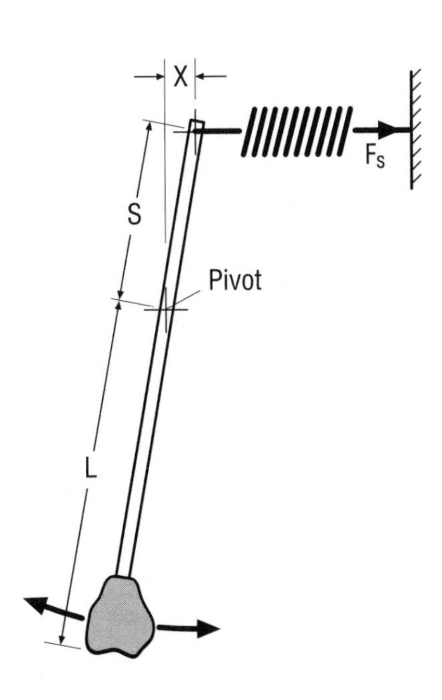

Try making the bob bigger and smaller with the same length of rod and note the time for fifty swings. Now try attaching the bob closer to the pivot and note the time for fifty swings. Try applying slightly more weight (if weight driven) or changing to a smaller resistor (if electric driven). How much does the pendulum period vary with each variable?

You can add a dial train and hands to the clock simply by arranging a pulley to rotate on an axle made from a nail or stiff wire in the center of the cardboard face as shown in the diagram. An elastic band connects the pulley to the motor shaft, and a cardboard pointer can be used for the second hand. What time does one rotation of the hand indicate?

THE SCIENCE AND THE MATH

If the pendulum swings only through small angles α, then it will feel an effective restoring force $mg \sin \alpha$, which at small angles is approximately equal to α in radians. The pendulum will be accelerated at a rate given by the familiar equation

Acceleration = Force/Mass.

In this case the acceleration

$$Ld^2\alpha/dt^2 = (-mg \sin \alpha)/m \sim -mg\alpha/m = -g\alpha,$$

where L is the length of the pendulum rod, m the mass of the pendulum bob, and g the acceleration due to gravity. (The minus sign appears because the force is in the opposite direction to the direction of increase of angle α). So the equation becomes

$$Ld^2\alpha/dt^2 = -g\alpha,$$

which is a type of differential equation that gives rise to solutions of form:

$$\alpha = \alpha_0 \sin \omega t,$$

where $\omega = 2\pi f$, in which α_0 is the amplitude of the oscillation and f the frequency of the oscillation. Hence

$$f = \sqrt{g/L}/2\pi.$$

With a length of 0.3 m, the pendulum will have a frequency of 0.9 Hz.

The pendulum frequency is not the only resonant frequency in the case of the crank pendulum. Another resonant frequency is the high-speed movement of the pendulum bouncing back and forth, driven by the spring. The simple case of a mass bouncing up and down on a spring is dealt with in all elementary texts on mechanics: if the force F due to the spring depends on extension as

$$F = -kX,$$

where X is the displacement from the mean position (k is a constant), then

$$-kX = Md^2X/dt^2,$$

which will yield oscillations of frequency f,

$$f = \sqrt{k/M}/2\pi.$$

Here the same equations can be applied, provided the force produced by the spring is considered to have been levered down to a lower value k', that is,

$k' = S/L.$

Putting a few numbers into these equations, you see that the characteristic frequency of the spring/mass resonance lies well above the gravity pendulum frequency. For example, with $M = 25$ g, $k = 100$ Nm – 1, and $S/L = 0.2$, you have

$f = 5$ Hz.

However, a mode like this could be excited if a large enough propulsion force is applied to the system, by giving more power to the motor or a larger weight on the drum, and the right starting conditions.

The system is complicated in practice by other factors, such as the weight of the pendulum rod and the flexibility of the rod, and there may be other resonant frequencies. By applying increasing power and noting the frequency of operation, it may be possible to distinguish other resonances of this sort.

And Finally, for Advanced Users

You can make a much slower pendulum without making the stick excessively long by making a compound pendulum, which means one in which there are more massive parts than the single bob of the simple pendulum. If, for example, another bob is added to the end of the pendulum rod that projects *above* the pivot (but not so heavy a bob or so long a rod that the pendulum does not hang down, of course), this increases the rotational inertia and decreases the restoring force, lengthening the swing time of the pendulum.

In the weight-driven version of the clock, a rather small-diameter shaft or drum is needed, and this has to have many turns of string, which can get tangled. You may find it convenient with weight drive to use two weights pulling in opposite directions, the larger weight rotating the shaft and pulling the smaller one up. The larger weight provides the drive as before, while the second weight simply ensures that the string grips the shaft securely and takes up the slack string, and avoids having as many turns of string around the shaft. With a secondary weight of one-tenth of the drive weight, you will need to wrap the string around the shaft only a couple of times. You will also find with this arrangement that you can rewind the clock easily by simply raising the heavier weight in the palm of your hand.

Readers who have tried a weight-driven version of the pendulum clock will have noted that the weight proceeds downward somewhat jerkily. Could you smooth out the motion of the clock by using a flywheel? How big a flywheel is needed? But would a flywheel keep the pendulum regulation from working?

16 *A Symphony of Siphons*

Blest pair of Siphons, pledges of Heaven's joy, Sphere-born harmonious sisters.

—With apologies to John Milton, "At a Solemn Music"

The ancient Greeks used water clocks called clepsydra, which were based on tapering pottery vessels with a small hole in the bottom. As the water ran out of the hole, the water level sank, uncovering graduations indicating the time. Clepsydra were used in court cases 3,000 years ago, presumably to keep the lawyers from arguing for too long. Evidently ancient Greece was not so very different from our modern world!

A cylindrical container with a hole in the bottom is a nonlinear water clock, since as the water level in the cylinder falls, the pressure driving the water out through the hole also falls, slowing the rate of water dribbling out. This can be compensated for by marking the time intervals in a matching, nonlinear way. Alternatively, making the vessel wider at the top causes the level to fall in a more linear way. With a linear fall in level, a clock dial mechanism (string attached to a cork float on the top of the water, with the string wrapped around a drum with a rotating hand) can be usefully linked to the clepsydrum. Mechanisms of this sort were brought to a fine pitch of ingenuity, being used to drive whole puppet theaters and other amusements, and are among the first complex machines ever made. Our digital siphoning clock thus follows in illustrious ancient footsteps.

A siphon refers to a tube that carries a liquid between a higher tank and a lower tank in which the tube goes up beyond the level of the surface level in the higher tank before descending to the lower tank. The descending portion is sometimes dubbed the tailpipe. The siphon can be used in a number of ingenious devices. The slow filling and rapid emptying of a siphon from a tank is most familiar from that everyday object, the water-flushed toilet (although there are designs that do not employ a siphon). This works by storing up a substantial volume of water (5 liters or so) in a vessel or cistern, and releasing this in a rush via a siphon; a lever that operates a piston or bell chamber allows the user to fill the siphon to start the action. As the basis for the clock in this project, we use a cascaded version of this action—apocryphally developed for the toilet by Thomas Crapper in 1886.*

The Degree of Difficulty

This project is easy—none of the dimensions needs to be exact, and no great quality of workmanship is needed. The complete lack of moving mechanical parts also promises reliability, potentially, as there is nothing to wear out. (Only a few years ago, manufacturers of electronic gadgets would proudly declare that their equipment was "all solid state," the implication being that there were no mechanical moving parts to go wrong inside. They don't use that selling line much now, but it remains true that equipment based on electronics without moving parts often is more reliable.)

What You Need

- ❑ Plastic vessels of increasing size (coffee cups, bottoms of 1 liter soda bottles, bottoms of 3 liter soda bottles, and so on)
- ❑ Pipes and elbows (copper or plastic, which can be easily attached to each other) or large-diameter bendable drinking straws
- ❑ Steady variable water supply (Some faucets are difficult to adjust to run at a constant, low flow rate.)

What You Do

Cut off the tops of the soda bottles. Make holes in the bottles that the pipes can fit snugly into, as the diagram shows.

*Thomas Crapper took a number of plumbing patents in the 1880s and 1890s, though none for siphonic flush toilets. These had already been developed at least a century earlier, as least in cruder form, according to Wallace Reyburn in *Flushed with Pride*. George Crapper, Thomas's nephew, did get an 1897 patent for an improved siphonic toilet, however. Despite this, Thomas Crapper's name is indelibly associated with the toilet.

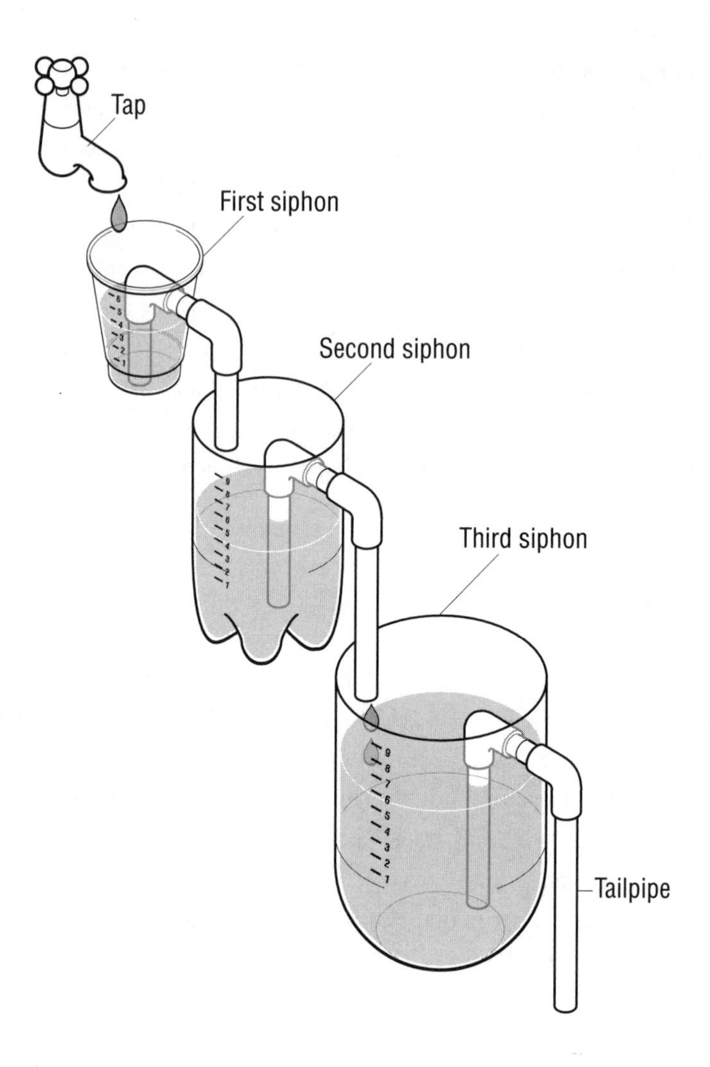

The ratio of vessel sizes should probably not be too great, or it will be difficult to empty the larger vessel before the next gush of water from the preceding vessel. I found that I could get ratios of around 10:1 easily. It might be convenient of course to use a ratio of 60:1 to give a conventional clock readout. However, this will be tricky. Also, more than twenty or so graduations on the water levels will be difficult to distinguish reliably using smaller vessels, and accurate digital readout could break down.

Our water clock is a digital version of the clepsydrum. In operation, the first small vessel (the oscillator) fills and empties every few seconds; the others, constituting the counting mechanism, take longer. So the second siphon empties every minute or two, the third every few minutes. Time can be read on the system in the same way as a clock is read in seconds, minutes, and hours, by reading the levels in the different vessels.

The Tricky Parts

You need to assure that your water supply is running at a suitable rate. If the water runs too fast, when the first siphon fills, water will overflow down the siphon continuously without the level dropping as it should. Too slow a rate, and fast dribbling sets in: the siphon never fills with water. Fast dribbling can also set in when there are air leaks in the siphon tube.

You must have an air gap between the tailpipe and the container it feeds. If you don't, the siphon will tend to remain full, rather than emptying neatly. It also appears that you must have a long enough tailpipe. If the tailpipe is too short, despite the resistance to flow being smaller, the siphon action is also difficult to set up.

The Surprising Parts

The siphon, when it triggers, runs quite fast, easily fast enough to beat the steady water input. However, it is not completely predictable. Sometimes the siphon will not trigger for several seconds: there may well be some intrinsic chaotic process involved in the formation. With a steady water supply, the clock can be surprisingly accurate over longer periods. The unsteadiness of the primary siphon is averaged out over a substantial number of cycles by the time the third siphon is reached.

Using Your Siphonic Clock

The flow rate of water can be varied over quite a wide range, and it is surprising how wide a range this is with some siphons. However, the siphon action stops only below a certain critical flow rate (you can get continuous burbling flow or a continuous stream down one side of the pipe).

The accuracy of the siphonic clock increases if the next-size siphon empties in less time than the cycle time of the smaller previous siphon. A high previous-siphon duty cycle is obviously helpful—this would also appear to suggest the need for a larger-bore pipe in a larger siphon.*

*The term "duty cycle" is often used in industry, and is applied to manufacturing machinery that is used in a regular cycle, for example. It is usually defined to be the amount of time the machinery is operating divided by the total cycle time. The relevant duty cycle here is the time taken to fill divided by the cycle time.

Again, for improved accuracy, there seems to be an advantage in using a larger pipe: stopping and starting is reliable, and the duty cycle is high. The slope of the longer down-going pipe of the siphon, as well as its length, seems to matter: very shallow slopes allow the pipe to fill easily but also allow excessive dribbling. I found steeper slopes good (around 20 degrees to the vertical).

Air in the water from the source may be a problem in some areas: air bubbles tend to appear when the air pressure above the water is reduced, which it is at the top inside the siphon tube. The bubbles cause trouble with large siphons that have tailpipes several feet long. There should be no problem with the device built as suggested, with maximum siphon height only the height of a plastic soda bottle.

THE SCIENCE AND THE MATH

As the water level builds, water begins to trickle over the top of the tubing elbows. However, the driving force for this flow is simply the excess of the height of the water over the elbows. The siphoning action begins when water fills up the siphon down-tube fully (however momentarily), at some point (forming a small "plug" of water). Once this has happened, the plug, still accelerating under gravity, forms a slightly lower pressure behind it in the top of the tubing. Water is thereby pushed at an increasing flow rate up the tube by the external atmospheric pressure, and this water forms further plugs, which in turn cause a further decrease in internal air pressure and a further increase in water pressure. Within a fraction of a second, typically, water fills the whole pipe and flows continuously. There is then a driving pressure ΔP equal to the depth ΔH of water times its density ρ:

$$\Delta P = \rho g \Delta H.$$

Note that the operative head of water, ΔH, is the distance from the water surface in the higher vessel to the bottom of the siphon outlet pipe, assuming the siphon to be at least momentarily full. With this strong driving pressure, the flow of water is now much more rapid than it was when just dribbling over the top, and at this flow rate the vessel is soon emptied.

The flow of liquid around a siphon is at first sight rather surprising: water has to flow uphill against gravity in the first part of a siphon. Only after this initial hill is climbed can the water be pulled downward by gravity in the more usual way. However, provided pressures do not fall too low, there is no reason why the hydraulic pressure in the liquid should not fall below that of the atmosphere. The positive pressure from the atmosphere on the surface of the liquid ensures that absolute pressure in the liquid typically stays positive. If the pressure in the liquid falls too low, gases dissolved in the liquid, such as oxygen and nitrogen from the air, will come out of solution and form bubbles, although not sufficiently to stop siphon action at low siphoning heights. Clearly, if the siphon is required to conduct liquid up a very long vertical column so that the absolute pressure falls below the vapor pressure of the liquid, there will be a tendency for the liquid to boil. Water might show this in a siphon more than about 10 m high, for example.

However, boiling is often delayed by the phenomenon of superheating. Superheating is where, in clean liquids, boiling occurs only when the external pressure is quite a lot lower than the vapor pressure. With or without superheating, if boiling starts, the siphoning action will stop, the column will fill with vapor, and the liquid will fall back into the vessel.

There are instances of siphon action at heights such that absolute pressure in the liquid is negative. These are seen in ultraclean laboratory glassware using specially degassed liquids with a tensile strength that holds them together. Water from the kitchen faucet in ordinary tubing and vessels does not show these phenomena, however.

And Finally, for Advanced Users

You can try different shapes of siphon pipe. As mentioned earlier, you can vary the angle to the vertical of the main tailpipe, but what would be the effect of bends in the long tailpipe? Would these allow the siphon action to form and the dribbling to stop more easily?

Try adding a little detergent or soap solution to the water feeding the clock. These substances are powerful surfactants that change the surface tension in the liquid by a large factor and often allow the buildup of bubbles too. What happens? Why? Presumably, lack of surface tension doesn't allow buildup of the blocking "plug" that fills the tube to create the rapid siphon action. If you use clear tubing for the siphons, you will be able to see what is happening.

To make the clock more like a mechanical clock, you could add floats attached to indicators: as mentioned, the Greeks did this with their clepsydra.

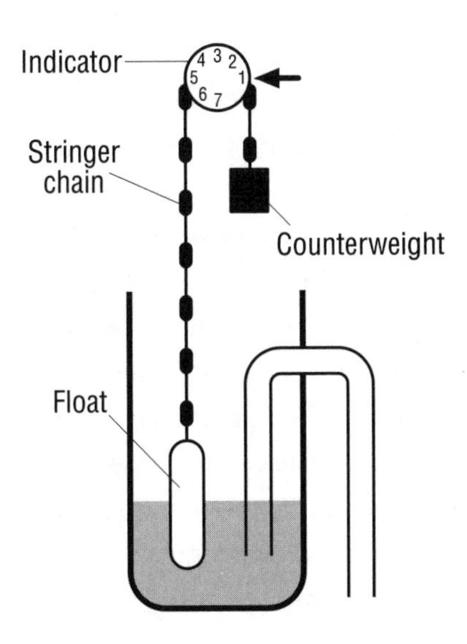

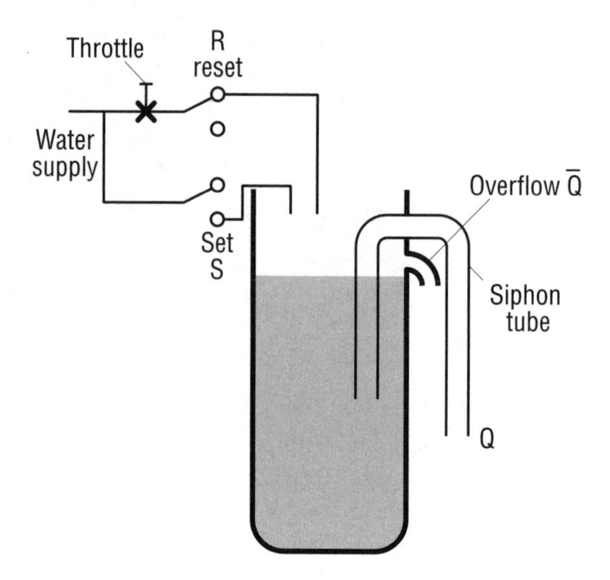

You could make a binary set of siphons, that is, capacities of successive vessels in the ratio of 1:2:4:8:16, and so on. This will need a fair bit of construction, but with such a small ratio between successive siphons, binary siphons may keep better time.

A siphon can be employed as part of a solid-state (no moving parts) dynamic flowing bistable device. A siphon and a simple overflow are the two outputs Q and not-Q or $\bar{Q}$ of the bistable vessel shown in the diagram. Normally, bistable circuits, whether electric or hydraulic, are symmetrical. Our bistable here is more like the asymmetrical bistable you might make, for example, with a semiconductor device like a tunnel diode.

A steady, continuous flow of water passes into the vessel at the top, carefully controlled in flow rate with the throttle valve (which could be the faucet valve or a piece of crimped rubber tube). This flow of water normally pours directly out of the simple overflow, which is halfway up the siphon outlet tube. However, if temporary extra flow (set or S input, in the diagram) is added that takes the water level temporarily up to the level of the siphon top, the siphon will start to work. It will quickly drain the vessel to below the level of the overflow pipe, and water (given the correct input flow rate and siphon diameter) will continue to flow down the overflow output. The vessel is reset (R input) by interrupting the steady input flow temporarily. This allows the siphon to drain out completely, so that when the input flow is restored, water goes down the overflow output again. This is not quite as easy to set up as it sounds, however—you have been warned!

Finally, if you want to know more about siphon flow, a lot of work has been published by hydraulic engineers; see B. H. Rofe's "Water Engineering," which refers to a whole symposium on the subject in 1976. There are descriptions of lots of different types of siphons used in hydraulic engineering. Rofe also gives details of siphon priming, where a small pump, perhaps as simple as a rubber squeeze-bulb for a laboratory siphon, is employed to initially fill the siphon and switched off once siphoning has begun.

REFERENCES

Reyburn, Wallace. *Flushed with Pride.* London: Trafalgar Square, 1998.

Rofe, B. H. "Water Engineering." Section B6, *Kempe's Engineers Yearbook.* 101st ed. Tonbridge, UK: Miller Freeman 1996.

Thorpe, Nick, and Peter James. *Ancient Inventions.* London: Michael O'Maran Books, 1995.

Whalley, P. B. *Two-Phase Flow and Heat Transfer.* Oxford: Oxford University Press, 1996.

17 *Bernoulli's Clock*

In Italy for 30 years under the Borgias they had warfare, terror, murder, bloodshed—they produced Michelangelo, Leonardo da Vinci, and the Renaissance. In Switzerland they had brotherly love, 500 years of democracy and peace, and what did they produce? The cuckoo clock.

—Orson Welles, in *The Third Man*

In his final speech in *The Third Man,* Orson Welles forgot to mention something else everybody knows: that the Swiss invented chocolate as well as cuckoo clocks.

All this is, of course, highly inaccurate. Cuckoo clocks came first from the Schwartzwald (Black Forest) region, which is in Germany, not Switzerland, around 1730, and chocolate was developed by the Aztecs of Mexico well before Columbus.

If they didn't invent chocolate or cuckoo clocks, what have the Swiss invented? Well, one thing that did come out of Switzerland (and still does) was good physics. The Bernoulli family was part of this for many years. Daniel Bernoulli, for example, developed the mathematics of flowing liquids and gases. In his book *Hydrodynamica,* published in 1738, he showed how the flow of liquids or gases can be explained in terms of the impact of microscopic atoms. Later, he worked on measuring force between electric charges and showed similarity with the same inverse square law that governs the gravitational forces between the planets and stars, again connecting the small scale with the large scale (work done accurately by Charles-Augustin Coulomb thirty years later using his torsion balance). He apparently did not make clocks—which is why it is left to us to make a clock using his equation.

In this clock, we utilize a curious prediction of the flow equation known as Bernoulli's equation: when you force a fluid to flow through a constriction, the pressure in the fluid will drop.*

Bernoulli's equation seems counterintuitive at first: if you squeeze a fluid through a narrow gap, then surely there must be a higher pressure in the gap? No! If a fluid is flowing at the same mass flow rate down a pipe, then the fluid must be traveling faster within the constriction than elsewhere, assuming the fluid is not accumulating anywhere but is flowing steadily. In order to make that fluid accelerate up to that faster speed, there must be a driving force; this can only come from a driving pressure gradient, which must be "downhill," that is, there must be a higher pressure before the constriction than within it. More counterintuitively, perhaps, the opposite must apply downstream of a smooth constriction: in order to slow down the fluid stream, there must a higher pressure downstream of the constriction.

The Degree of Difficulty

This is a more difficult project. You need a good air blower and a stand for the pendulum that has an adjustable mounting for the plate, so there is a little engineering to be done.

What You Need

- ❑ Pendulum stick and clay weight
- ❑ Nail for pendulum pivot
- ❑ A flat plate (such as a 60 mm diameter × 1 mm plastic sheet)
- ❑ Air blower†
- ❑ Plate with a hole in it to fit the blower nozzle (optional)
- ❑ Low-voltage power supply for the air jet, if required
- ❑ Small-diameter drum (for second hand; I used a thread spool about 5 mm in diameter)
- ❑ Plastic straw (for second hand)

*Bernoulli's famous equation is a compulsory section in every major textbook that deals with gas or liquid flow. Chemical engineering texts like J. M. Coulsdon and J. F. Richardson's *Chemical Engineering* (pages 27–35) have it, every aeronautical engineering book has it, and even my own book on industrial gases has it.

†An electric air-mattress or air-toy inflator works; these are small centrifugal fans than run off 12 V and cost around $15. A hose attached to the exhaust outlet of a vacuum cleaner set to low speed will also do, or any other source of an air jet similar in strength to that of a person blowing. A hair dryer with a cold setting also might do, although beware here, because many hair dryers blow rather warm air even on the cold setting, since they use the heater windings to reduce the domestic main voltage down to that needed by the fan motor.

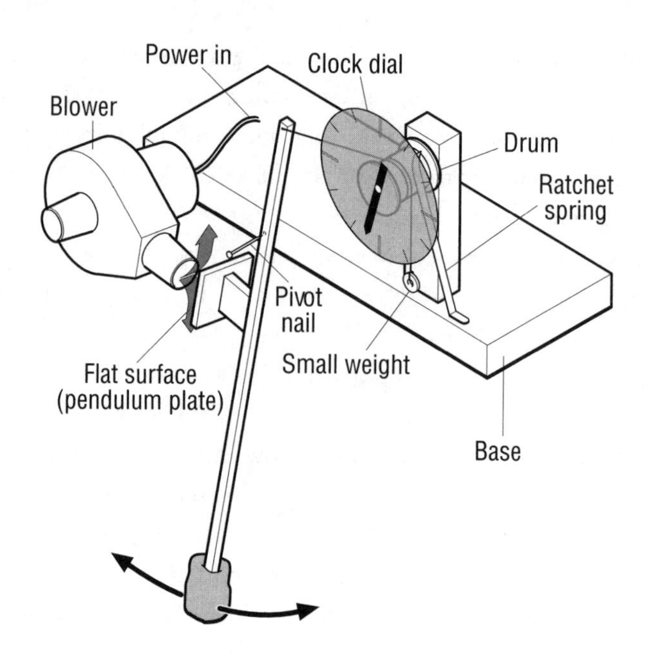

- ❏ Small weight (such as a metal washer)
- ❏ String (I used extra-strong thread.)
- ❏ Wood for mounting the hardware

What You Do

Erect a freely swinging pendulum first, and then fit the flat surface—the pendulum plate—for the applied power from the air jet (see the diagram). This constitutes the oscillator and "going train." Don't attach the string that will go from the pendulum to the "dial train," the part of the clock that counts and displays the time yet—do that later, after the pendulum is working. The air jet must then be positioned to blow on the flat surface on the pendulum. The set-up must be arranged so that the air jet contacts the pendulum plate only when the pendulum has begun its upward climb on the near side. If the air jet touches the pendulum plate too early, then the device either will fail to work or will have a very restricted swing.

The drum must be lightly retained in its position by friction. I used a light spring pressing against the edge of the drum.

Switch on the air jet, swing the pendulum, and adjust the position of the pendulum plate and jet. Then try again . . . and again . . . until you get sustained oscillations.

Set the pendulum swinging. When the plate approaches the jet, it will be *sucked* in toward it, and then pushed away. It is sucked in because of the magic of the Bernoulli effect.

But there is another surprise: the pendulum is not sucked in and then stuck on the end of the jet like an insect on flypaper. The Bernoulli effect occurs only when the flow is rapid, and the suction disappears along with the flow when the end of the jet is blocked by the flat surface and the flow speed reduces. Furthermore, there is no equilibrium position: the pendulum does not simply hover a millimeter or two from the end of the air jet but soon starts to jig back and forth and then swings steadily.

Once the pendulum is swinging steadily, wrap the string from the pendulum around the drum about 1¼ times and attach a small weight to its end. With a suitable weight and adjustment of the drum friction spring, each "back" swing of the pendulum will pull the drum along one increment, the string tightening on the drum and pulling it around. On each "forward" swing of the pendulum, the string is slacked off and allows the weight to drop back, leaving the drum stationary—without rotating the drum backward.

THE SCIENCE AND THE MATH

The Bernoulli equation can be summarized as

$$P_i + \tfrac{1}{2}\rho U_i^2 = P_j + \tfrac{1}{2}\rho U_j^2,$$

where P_i and P_j are the pressures at two points along the pipe, ρ the (constant) density at those points, and U_i and U_j the fluid velocities at those points.

So if a smooth constriction in the pipe increases U_i above U_j, then the pressure at J, P_j, must drop below P_i. The equation follows quite naturally by considering the energy of an element of the fluid flowing in a pipe. An element of mass M naturally has a kinetic energy of $\tfrac{1}{2}MU^2$. But the mass M per unit volume is simply ρ, so the kinetic energy of the fluid element per unit volume is simply $\tfrac{1}{2}\rho U^2$.

The other element of the equation relates to the energy of the fluid simply because of its internal pressure. Imagine that an element of fluid (a kind of "gaseous toothpaste" may be easier to visualize) is squeezed out into the vacuum of outer space, just for a moment. If a volume V is emitted by a hole of area A, then the length of the toothpaste emitted will be L, where $V = LA$. Furthermore, the force acting on the toothpaste from the inside will be PA, where P is the internal pressure. An amount of work equal to force × distance = PAL will be done on the toothpaste. This energy per unit volume due to pressure is then simply $PAL/V = PAL/AL = P$, which is what appears in the Bernoulli equation.

In this case, the air flow coming out of the blower nozzle and over the flat plate that nearly blocks it is the constriction on which the Bernoulli equation applies. The nozzle is found to push the plate away when the spacing is large, as might be expected, but is found to *attract* the plate at short distances. The constriction of flow forces the speed of the air to increase, and the pressure to decrease, in accordance with the equation. If the plate is just 1 mm away from the nozzle of, say, 12 mm diameter, then the area of flow decreases from 113 to 38 mm^2. With a static pressure of, say, 0.002 atmospheres on the inlet, the pressure will drop to around –0.002 atmospheres on the outlet at the flow rate I had with my pump. (I used an inflator pump with a capacity of about 50 l/m peak, although this was changed a great deal with the proximity of the plate.)

The behavior of a plate with a nozzle blowing air against it is of course more complex than this, overall. At long distances from the plate, the air from the nozzle pushes it away—visualize the air as a stream of billiard balls bouncing off the plate. It is only when the plate is quite close (a few millimeters, for a 25 mm diameter nozzle) that attraction occurs due to the Bernoulli effect. Furthermore, that attraction is reversed again to become a repulsion at very small (submillimeter) distances. At those distances, the flow rate is so small that the pressure on the plate simply becomes the static positive pressure of the output of the blower. The graph gives an idea of these effects with the blower I used, connected to a 60 mm diameter circular plate above a blank plate. The distance of the blank plate from the blower plate is plotted versus the force on the lower plate.

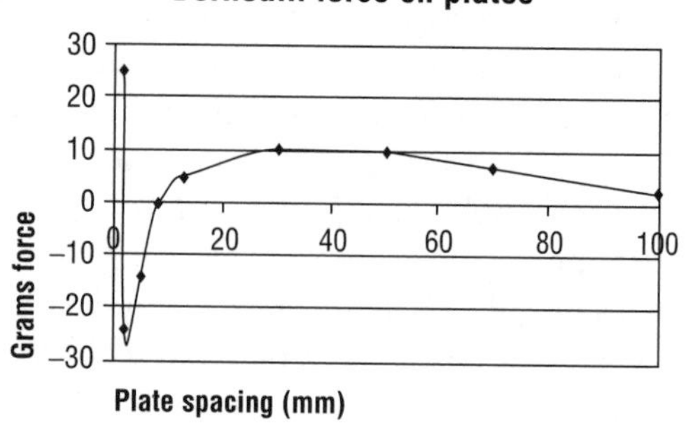

There are also simple equations governing the behavior of the "rope ratchet" dial train (here, it is more appropriate to refer to it as a "string ratchet"). Each tiny segment of the drum around which the string is wrapped can be considered to behave as in the drum diagram. The differential equation governing the segment is

Normal Force = $2T \sin d\alpha$

Frictional Force = μ

Normal Force = $2\mu T \sin d\alpha \sim 2\mu T \sin d\alpha$
(for small $d\alpha$),

where μ is the coefficient of friction, T the tension in the string, and $d\alpha$ the angle subtended by the segment of string.

Thus $dT = 2\mu T\, d\alpha$.

This is simply integrated to give $T = \exp(2\mu\alpha)$, or $T = K^n$, where K is a constant and n the number of turns of string on the drum.

This equation has a number of interesting properties. First, its exponential form indicates how each turn of the string on the drum gives another factor of K in the equation, meaning that an input force of 1 unit is multiplied by K on the output with one turn. With two turns, the input force of 1 unit is multiplied by $K \times K$, that is, K^2. Second, the constant factors entered into the equation are the key to why it works as a ratchet.

When you pull on the string, it will grip with a force of $K^n Mg$, where n is the number of turns on the drum, M the suspended mass, and g the acceleration due to gravity. This is so large that it will cause the string to grip the drum tightly and rotate it against the friction device, pulling the weight up a little. When you "push"—that is, let the string go—the mass M will exert a force Mg, pulling it back to its starting position, and the string will not grip the drum. This is because—although the drum is multiplying the input force by K^n—this time the input force is zero, and multiplying it by an even larger factor will still give zero, and hence the drum will not be gripped and the weight can drop freely.

And Finally, More Bernoulli Clocks

Clearly the rope ratchet can be used for other applications. If you want to control the movement of a large object, you can use the friction of the rope wrapped on the drum to magnify your own effort. Suppose the multiplication factor K (see "The Science and the Math" section) is 10× per turn on the drum, and you want to control the movement of a large boat buffeted by waves at the dock. If the ship weighs 10 tons, then it might pull with a force of 1 ton as it moves with the waves. Then if you can pull with a force of 10 kg or 20 lb, you can control the ship with just two turns of the rope on the drum. If the ship was ten times bigger, then (provided the rope is strong enough) you can simply control this bigger ship with a third turn on the drum.

You can of course attach a more elaborate going train to the clock: with pulleys or gears, you can add a minute hand and even an hour hand. One of the difficulties of calibration is that even with the pendulum keeping perfect time, the gear train is relying on the rope ratchet. And the rope ratchet does not itself move through a perfectly defined angle for each pendulum swing in the way that the escapement wheel of a clock turns exactly one gear tooth for each swing. This difficulty could be avoided by using a more conventional gearwheel and a

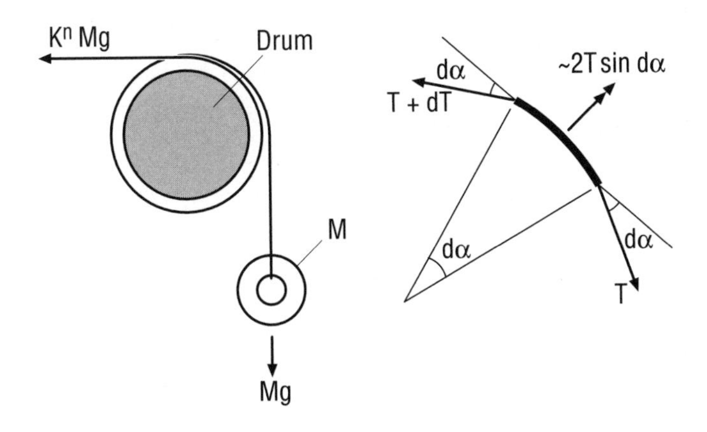

driving pawl, but this is more difficult to engineer, as well as less interesting mathematically!

REFERENCE

Coulsdon, J. M., and J. F. Richardson. *Chemical Engineering.* Rev. 2d ed. Oxford: Pergamon Press, 1965.

Curious Conveyances

18 *Dougall or Vibrocraft*

There are two sorts of technology: exponentially developing technology and unimportant technology.

—Ian Ross, president, Bell Laboratories

If Ian Ross was right, and the only important technology is exponentially developing, I guess the Dougall is an unimportant technology. Only if this book is a rip-roaring success and huge numbers of you go off and make Dougalls will the Dougall stand any chance of being other than unimportant. And strictly speaking, only if you then go and tell your friends to make some more Dougalls, and they in their turn tell their friends, and so on ad infinitum will the Dougall really meet Ian Ross's definition. Alternatively, a reader expert in genetic engineering might like to produce a self-reproducing Dougall.

It seems to me that technologists since about 1900 have been obsessed with rotating machinery. Ever since the rise to prominence of the turbine and the electric motor, machines with pure rotative motion, we have neglected to some extent the possibilities of reciprocating or vibrating machinery. Why otherwise did the motor industry spend so much effort on the Wankel rotary internal-combustion engine? And why else did Eric Laithwaite have to remind the world about the possibilities of linear electric motors in the 1950s? There are familiar examples of vibrating machinery like the vibratory bowl feeders that supply nuts and bolts from a bowl to a robot assembly machine, and there are much less familiar examples, like the extraordinary oscillatory compressors described in John A. C. Kentfield's book on pressure exchangers, *Nonsteady One-Dimensional, Internal, Compressible Flows*. But these successful examples are rare compared to the legion of pure rotative machines that surrounds us today.

I call vibration-driven vehicles that use brushes underneath *Dougalls* because they look and move rather like a long-haired sheepdog named Dougall who was one of the stars of a long-running French children's cartoon, "Magic Roundabout." Dougall looked a little like a hairy mop with a face on one end. A Dougall is of course very good at sweeping the floor as it goes, making it a highly environmentally friendly vehicle!

Vibrocraft are definitely not in the mainstream of vehicular invention. Nowhere in the almost boundless variety devised by natural selection has Nature come up with a Dougall. Even Meredith Thring, the eccentric London professor famous for his curious vehicles, including a walking farm tractor, has not built a Dougall. This is not to say that they have been entirely neglected, however. In Japan someone actually built a full-size Vibrocraft that used an array of slightly flexible plastic (round-ended) rods a few inches long to support both itself and the enthusiastic driver. The driver in the newsreel I saw looked like a cartoon of a pneumatic road-drill operator—he was being jiggled up and down too much for comfort. And various toys have been devised using vibration action (see

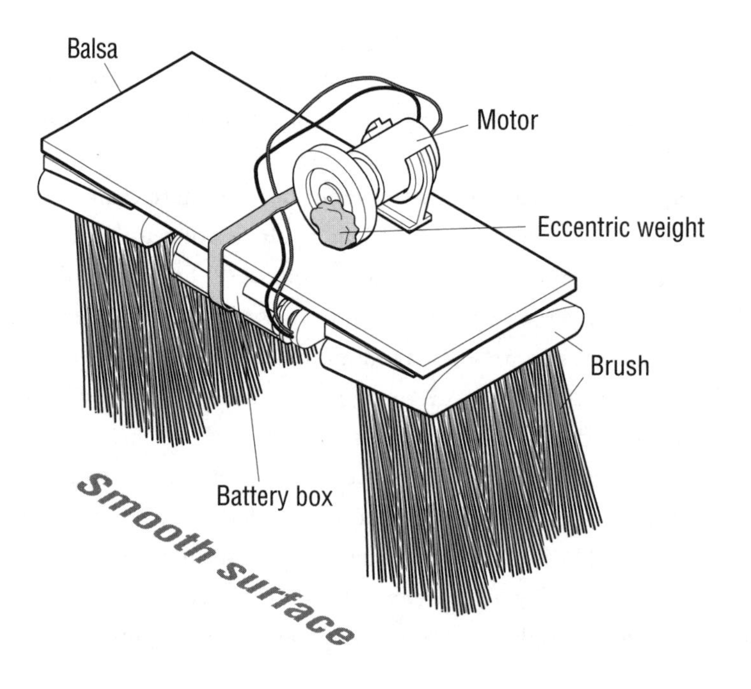

Melissa M. Truax, "Cat Toy"). Tiny clockwork-driven jiggle-bugs with long wire legs instead of brushes underneath, some beautifully crafted in stainless steel, are available in stores.

The Degree of Difficulty

This is basically an easy project, particularly if you happen to be a qualified hairdresser. There is really only one point of difficulty: the brush. Some brushes will need trimming, and trimming a brush to make the Dougall stand up straight and travel in a straight path can be tricky.

What You Need

- ❑ Broom head
- ❑ Small electric motor
- ❑ Battery box and rechargeable batteries (The Dougall uses a relatively high current in the motor and quickly exhausts batteries.)
- ❑ Wheel to fit the motor, 20–30 mm diameter
- ❑ Eccentric weight (such as a steel nut)
- ❑ Sheet of balsa wood
- ❑ Balsa wedges
- ❑ Glue and tape

What You Do

Cut two 25 mm slices from the broom, then glue these, perhaps using balsa wedges, to the balsa sheet, ensuring that the bristles are angled backward by 10–20 degrees. Add the motor, with the eccentric weight firmly glued or taped to the wheel. Insert the batteries into the battery box and firmly glue or tape it to the bottom of the balsa sheet.

The Tricky Parts

Make sure that the small cables from the battery to the motor are supported by tape so that they do not bend back and forth at the ends where they attach. Such bending will cause rapid fatigue failure of the copper wire inside the cable and stop your Dougall in its tracks.

If your broom has rather soft and pliable bristles, you may find that, even with careful balancing, the Dougall simply falls over. In that case you need to trim the bristles shorter. As noted, this is more difficult than it sounds, as it very easy to give your Dougall an uneven haircut, which will make matters worse.

The Surprising Parts

Why on Earth does a Dougall work? The answer is, it doesn't work on the Earth at all. At high power, at any rate, it makes little jumps upward and moves in the air! Actually, it doesn't have to leap into the air to move along, it can simply slide forward a little each time it moves its eccentric weight backward, and when the eccentric weight moves forward, the Dougall tends to slide not as much backward as it did forward.

Using Your Dougall

A Dougall will go up very gentle slopes and can sometimes be trained (by adjusting the position and orientation of the motor) to go in a straight line, although a circle is much more typical. By tethering the Dougall with a light thread to a central point, so that the thread is tugging in the opposite direction to the Dougall's natural turn, you can constrain the craft to run in precise circles, which is useful for measuring how fast it is going.

Try your Dougall out on a rough surface. Now try a rough and springy surface (carpet, for example).

THE SCIENCE AND THE MATH

Although the Dougall doesn't need to, if it leaps upward as it moves forward, even if it doesn't leap in the air but simply reduces the weight exerted on its bristles, it will tend to go faster. This is because the frictional force exerted by the brush bristles on the surface, which stops it from moving, is proportional to the downward force due to the Dougall's weight. The leap upward reduces this downward force and hence reduces the friction. The bristles of the brushes help to define the direction of motion, because they tend to bend backward, allowing the Dougall to fall slightly. This reduces momentarily the force of the bristles on the ground and allows easier movement forward when the Vibrocraft is moving forward. Conversely, when the Dougall tries to move backward, the bristles bend forward and try to raise the Dougall upward slightly, exerting a higher force on the ground and giving a higher frictional force. There may also be other less important effects due to the ends of the bristles having a different coefficient of friction than the sides.

In fact, if the motor is sufficiently enthusiastic, and the eccentric weight is heavy enough relative to the overall weight, then the Vibrocraft will actually take off. It is easy to calculate the takeoff speed. When the force due to gravity (g) on the whole craft (mass M) exceeds the force needed for the acceleration of the eccentric weight (mass m) going around at an angular speed of ω radians per second, that is, when $m\omega^2 r \sin \omega t > Mg$, or $\omega > \sqrt{Mg/(mr)}$, the Dougall will take off, at least for a part of the rotation of the eccentric. For example, with M/m at 50:1 (that is, a 100 g Dougall with a 2 g eccentric weight of 1 cm radius), a rotation of about 2,000 rpm will suffice.

You may find that you can excite a resonant frequency of the Dougall. By varying the power put out by the batteries, and the size and off-axis distance of the eccentric weight, you may find that you can get the Dougall to vibrate at its resonant frequency, giving a bigger amplitude of vibration and allowing it to go much faster. The mass of the Dougall body and the springiness of the bristles can be such that it will perform harmonic motion, although it is heavily damped. The resonant effect is not strong in the ones I have tried. You can easily see by pushing down and then letting go that the Dougall is highly damped. However, even though it is damped, there is a resonant frequency, and it will show some advantage when running at that frequency.

And Penultimately, for Advanced Dougall Users

You can make a Vibrocraft that goes faster than a Dougall by making the body out of a very lightweight expanded polystyrene, and giving it a "skirt" of some kind—bristles or perhaps just a rim of expanded polystyrene. You will probably find, however, that such a Vibrocraft has a tendency to move off sideways and to go around and around instead of forward and is generally a bit trickier to control than the broom-based Dougall. You could use a set of springy wire legs beneath the vehicle, perhaps with tiny rubber feet on their tips, which would give you a high-powered version of one of the popular clockwork jiggle-bugs. However, like the polystyrene Dougall, controlling the vehicle to go in a particular direction may prove impossible.

And Finally, a Very Advanced Radio-Controlled Dougall

You may have found that your Dougall goes forward and to the left with the motor running in one direction, and forward and to the right with the motor running in the other direction. By using a radio control set (perhaps removed from a simple and inexpensive forward/backward toy car), you could provide your Dougall with radio control. The motor on the Dougall is connected instead of the toy automobile's axle. By pressing the button intermittently on and off, you can make the Dougall go in a rather wiggly straight line, while holding down or releasing the button allows turns.

Will the Dougall go faster if a propeller is added? The low friction of the Dougall on the top of its vibration cycle may allow the weak thrust from a small propeller to push it forward, faster than it moves by vibration alone. Or will it?

REFERENCES

Kentfield, John A. C. *Nonsteady One-Dimensional, Internal, Compressible Flows.* New York: Oxford University Press, 1993.

Thring, Meredith W. *Man, Machines, and Tomorrow.* London: Routledge and Kegan Paul, 1973.

Truax, Melissa M. "Cat Toy." U.S. Patent #6155905. U.S. Patent and Trademark Office, Washington, D.C.

19 *Follow That Sunbeam*

> He had been eight years upon a project for extracting sunbeams out of cucumbers, which were to be put into vials hermetically sealed, and let out to warm the air in raw inclement summers.
>
> —Jonathan Swift, *Gulliver's Travels*

Jonathan Swift was remarkable in his anticipation of solar-power energy storage. But in the words that follow those quoted in the epigraph, he goes on from extracting sunbeams from cucumbers to further predictions, including science-funding crises of the future. "He told me he did not doubt that in eight years more he should be able to supply the gardens with sunshine at a reasonable rate; but he complained that his stock was low, and entreated me to give him something (money) as an encouragement to ingenuity, especially as this had been a very dear year for cucumbers." This must strike a chord with every scientist or engineer who has had to beg for funding for an R&D project.

Gulliver's Grand Academy of Lagado apparently did not attempt to employ sunbeams to guide missiles. But the principles of a homing device must be intuitive to anyone who has chased an animal or watched such a hunt. And gunpowder rockets were available long before cannons, so the idea of the homing missile could have been used by Swift.

The principles of the homing device or beam rider seem only to have arisen more recently, however. In the nineteenth century, "leading lights" began to be employed. These were themselves improved versions of "leading marks"—prominent features of the landscape, sometimes painted white to enhance visibility. Mariners in Europe could use two lights placed at different heights on the shore and aligned with a safe channel to guide them into harbor. The lights (and marks) still exist in many places, and I have used them myself at sea. By a reversal of this logic, the lights can be used to guide a boat out of the harbor; the boat is then said to "ride the beam."

Beam riding was first forged into a weapon of war with the radio beams of the Second World War, the German Knickebein system being the first to be used, steering manned bomber aircraft. Winston Churchill describes this system in "The Wizard War" in the second volume of his history of the war. He also describes the jamming signals that were used to counteract it, and he tells how on one occasion a whole cataract of bombs, hundreds of tons of them, was released by the marauding force on an unsuspecting—and entirely unoccupied—field north of London.

It was not long before these principles were applied to steerable missiles. Small-wheeled vehicles running around on the ground were used in most of the early experiments on guidance technology, and the more successful ideas were incorporated into crude missiles and robot aircraft by the end of the war. A postwar seminar described beam riders—TV control, radio and wire control, and even optical beam-riding systems that were developed by 1945 in Germany but not extensively deployed. However, it was not until the Cold War era of the 1950s that guided missiles really changed the face of warfare, perhaps for the better. For the first time, instead of firing hundreds of inaccurate unguided explosive shells, killing many innocent noncombatants and pulverizing the surroundings, a single guided missile could often incapacitate an enemy with a minimum of casualties.

With the aid of four electronic components, a couple of electric motors, and some ingenuity, you can make your own Guided Carpet Missile. You can simulate much of the Cold War smart-weapon ploys of measure and countermeasure on a rug. Beam-riding and homing modes are possible, and you can try jamming and decoys.

The Degree of Difficulty

This project does need at least basic skills in both mechanics and electronics, but it should not prove too difficult. The mechanical assembly—the transmissions—

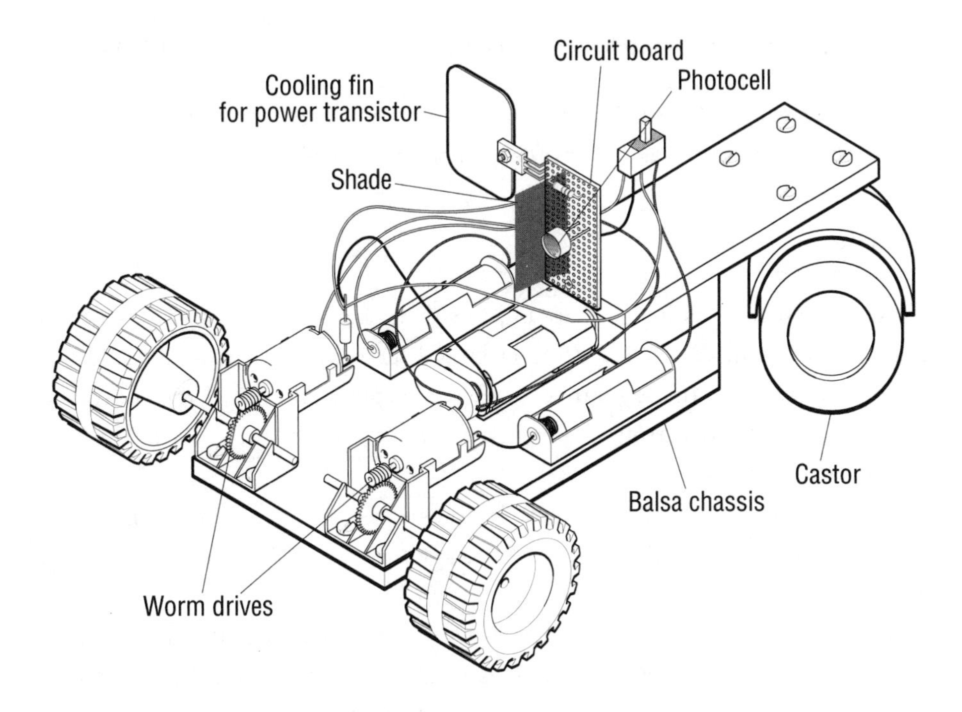

needs to work smoothly, and the electronics needs correct assembly. However, nonelectronics buffs may want to assemble the electronics, for example, using a screwdriver and screw terminal blocks, rather than more conventional soldering. New parts from suppliers, of course, have long leads that can be easily pushed into screw terminal blocks.

What You Need

- ❑ 2 electric motors (within the capacity of the batteries used, 1.5–6 V, and less than a couple of amps current)
- ❑ 2 transmissions (a simple single-stage worm drive may suffice) with a speed reduction of about 20:1*
- ❑ 2 wheels, 30–60 mm diameter
- ❑ Castor (from an office chair, for example)
- ❑ Wood for frame
- ❑ Batteries and battery boxes for electric motors (rechargeable batteries, at least AA)
- ❑ Transistor
- ❑ Power transistor
- ❑ Resistors

*If you can't buy these directly, even from a model shop, you may find it easier to dismantle some electric toys and use their transmission or motor units.

- ❑ Cadmium-sulfide photocell (I used one out of an old camera, but an ORP12 or others will do.)
- ❑ Matrix circuit board (small piece) or screw terminal "chocolate block"
- ❑ Flashlight or bicycle lamp with a defined narrow beam

What You Do

Build the assembly as in the diagram, but then change the battery voltages to ensure that one motor runs continuously and the other is switched on and off, so that the Carpet Missile follows a course sharply to the left if the controlled motor is off and less sharply in a rightward direction with the controlled motor on (or vice versa).

To make the photo-sensor work, it needs to be on the edge of a shadow—either of a specific shade, or perhaps just using the edge of the sensor itself or the edge of the beam.

For homing purposes, this edge is provided by a shadow that falls across the middle of the sensor when the vehicle is headed straight for the light source. If the vehicle turns slightly too far to the left of the direction of the homing beacon, then the shadow on the photocell lightens, and this speeds up the port side motor to move the vehicle to the right slightly—and vice versa if the vehicle strays too far to the right of the homing beacon.

The edge of the controlling light beam is used in the beam-riding mode. In beam-riding mode, with its sensor in the light beam, the starboard motor races and the Carpet Missile heads off on its curved track toward the left edge of the beam. When it meets the edge of the beam, the starboard motor slows and the vehicle turns back toward the right side of the beam, then meets full light-beam

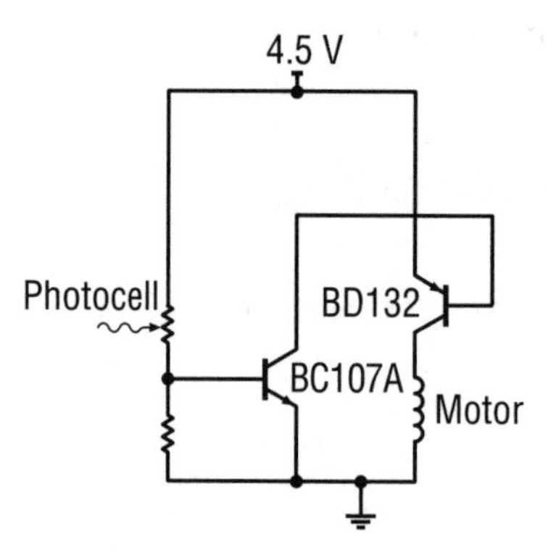

strength and wiggles back to the left again, in this way following the left-hand edge of the beam.

The Tricky Parts

The castor wheel must be free running and free swiveling (properties that may not be met by the brass castor off your antique grand piano). If your Carpet Missile turns around and around without going anywhere, you may find that the photocell circuit either is not working or is connected to the wrong motor.

The speed of the controlled motor should be at maximum about twice that of the continuously running motor. You can adjust either motor's speed by changing the number of battery cells supplying each one. You can also reduce either motor's speed by adding one or more power diodes (with sufficient current rating) in series with the motor: each diode will remove about 0.6 V from the voltage powering the motor (see "Useful Materials and Components" at the end of this book).

The Surprising Parts

At first sight the Carpet Missile might be expected to perform a violent wiggling dance—a high-powered prototype version did exactly that! However, the choice of relatively slow-responding motors and a chassis that acts to damp out rapid movements means that the Carpet Missile is capable of a relatively smooth trajectory, although it does hunt from side to side of the beam by a few centimeters.

Using the Guided Carpet Missile

The Carpet Missile has two basic modes of operation, depending on whether it is heading toward the light beam source (homing mode) or away from it (beam-rider mode), as mentioned. An ordinary table lamp suffices for the homing mode. However, for the beam-rider mode, you need to have a well-defined narrow beam such a bicycle lamp or flashlight. Because the missile will try to drive away from the light source along the edge of the beam, that edge needs to be well defined.

To operate the Carpet Missile, turn on only a dim ambient light in a dark room. Then try out your homing light or rider beam. You should find that the Guided Carpet Missile shuffles along, weaving from side to side, accelerating and decelerating as it goes, with the controlled motor alternately slowing and racing.

By using, for example, red-colored lamps in the room where you operate, and using a blue or green filter and guiding beam on the Carpet Missile, you can make it work under brighter illumination.

You can pass the Carpet Missile from beam to beam, if one person switches off his or her lamp or flashlight. Can you pass the machine around both a left-hand turn and a right-hand turn, or only a left-hand turn? Can you "steal" the Carpet Missile from its initially chosen beam by a suitably bright alternative source? And by using two lamps alternately switching on and off, can you decoy the missile in its homing mode into homing in on absolutely nothing?

THE SCIENCE AND THE MATH

The ORP12 photocell functions as a conductor when light is on it. Essentially, the light generates pairs of conductors, negative electrons and positive "holes," in the cadmium-sulfide semiconductor by splitting electrons from some of the atoms in the device. The electrons and holes created by the light make the device more conductive, and the small current that flows can switch on the two transistors. The two transistors make the assembly more sensitive and step up the current from a milliamp or two to the level at which it can operate the electric motor on 1 amp or so.

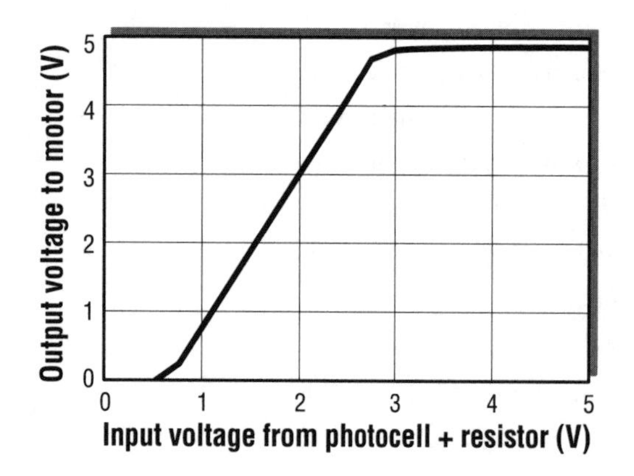

To make the vehicle home in on a source, if the photocell is on the port side, the photocell must drive the vehicle's starboard-side motor. In the homing mode, the vehicle homes in by orienting itself so that the shadow across the photocell tends to be kept at a constant angle to the light source. If the source does not move and the photocell shade is lined up accurately with the vehicle axis, then this will be a (more-or-less) straight line. More subtle, of course, are the questions of how the vehicle moves if the source moves, and how the vehicle moves if the shade on the photocell is significantly off the vehicle axis. (It may help to remember that when two vehicles are moving in straight lines on a collision course, the bearing of one relative to the other is constant.)

In beam-rider mode, with the photocell on the port side (facing backward now), the photocell must drive the vehicle's starboard side motor. The beam from an ordinary flashlight or bicycle lamp spreads out, and the edge of the beam will become fuzzy over a few meters. As the beam gets fuzzier, the control of the Carpet Missile gets less effective—you will find it tends to wiggle more from its path along the edge of the beam. You can redirect the vehicle somewhat by sweeping the beam slowly in a horizontal plane.

The exact path followed by a beam-rider system is a more difficult problem, particularly when the edge of the beam is significantly ill defined, and if the beam is being moved. The problem is similar, in principle, to many in control system theory. In industry, for example, the one-dimensional tracking task of making a furnace run at a constant temperature is related. This is normally solved by the use of a proportional, integral, and differential (PID) controller unit. This provides an output drive J, which is given by

$$J = AI + BdI/dt + C\int Idt,$$

where I is the input signal (normally the difference between the actual temperature and the desired temperature), and A, B, and C are constants. Essentially, the integral term prevents a permanent offset between desired and actual temperature (because it gets bigger and bigger if that offset remains), while the proportional term provides the basic corrective action. The differential term accelerates the response to a sudden step change in the input. The constants are chosen to optimize the response—to give a reasonably quick response without too much overshoot and without too long a delay in correcting an offset.

The extension of this to two or three dimensions is possible, and there are whole books on control theory that cover this, mostly heavyweights aimed only at practitioners. However, William Gosling's *Helmsmen and Heroes* is more readable and applies the control theory not just to mechanisms but to all manner of systems, from pig farming to politics.

I was once called upon to find out why a certain model of large industrial boiler operated correctly only at constant firing and was subject to wild fluctuations and occasional minor explosions during change of firing. It turned out that the two wires had been swapped, changing the sign of the constant B in the equation when the circuit diagram was translated into copper tracks on the printed circuit board used in the boiler controller. The untimely death of the designer meant that this fault had gone into the production equipment undetected and made the problems doubly enigmatic. However, once the mystery was uncovered, a simple swap back of the two connections and the boiler worked much better!

And Finally, for Advanced Users

As noted, the requirements to make the Carpet Missile operate in beam-rider mode are slightly different. Here the missile must traverse smoothly across the floor until it intercepts the edge of the beam, all the time progressing away from the light source. It would clearly be an advantage to use a much better-defined beam, and suitable modifications (perhaps a different photo-sensor with a red filter) would allow the use of a laser beam from a laser pointer, for example, as the guide beam.

For the electronically minded, it may well be possible to improve enormously the path followed by the Carpet Missile by the use of PID techniques (see the previous section). The output from the photocell is used as the control signal, and the controlled motor is connected to the output of the PID system. PID controllers can be purchased as units or constructed from Op amps. Classic electronic texts like Winfield Hill and Paul Horowitz's *The Art of Electronics* describe how this is done. However, perhaps simpler devices might improve the response of the device: adding a small flywheel, for example, to smooth the on/off nature of the controlled motor's response.

Real guided weapons frequently use yet another mode: target reflective mode. This is where an illuminated spot on the target acts as the homing source. The problem with this set-up is that the sensitivity of the sensor needs to be much higher: normally a modulated beam is used, and the sensor is tuned to that modulation. For our device, the sensitivity is typically limited by the ambient light. An increased-sensitivity sensor can only be used if the missile is operated in nearly total darkness. However, perhaps the ingenious reader can devise a simple way of achieving ambient-light rejection.

The missile could perhaps be persuaded to follow a more complex path than the relatively straight line it follows with a fixed beam or homing beacon, if the light beam it followed could be bent around corners. A set of two or three mirrors could be used to provide an outward and return path, for example, with the principal problem being how to follow the beam without the machine colliding with the mirrors.

The Carpet Missile tends to describe a track roughly in arcs of circles—radius r to one side and radius $2r$ to the other, modulation of speed from about $0.25v$ to $0.75v$ synchronized with the changes of course, where v is the speed given by the speed of the faster motor. As more battery voltage is added (or subtracted by dying batteries!), you can measure excursions to either side of the main path, and you can see the effect of the shadows cast by different photocell shrouds.

There is also a looping homing track that the Carpet Missile can follow. Wherever the photo-switched motor is not quite powerful enough to overtake

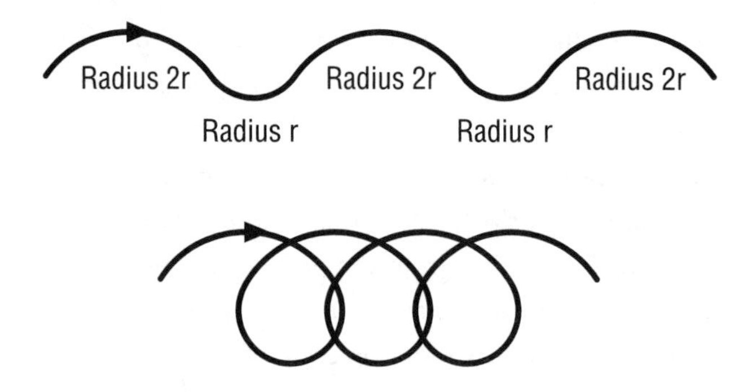

the motor, which is normally on, the Carpet Missile will follow a gentle curve toward the lamp. It gradually moves its sensor away from its shade, until the edge of the sensor, and its nonorthogonal angle to the homing lamp, is such that the photo-switched motor turns off. Then the Carpet Missile turns sharply around until the shade uncovers the photo-resistor again, when the rate of turn once again drops to a very slow progress.

The resulting orbit is not a gentle undulating track with opposite turn directions as described earlier, but a kind of cycloid with the machine always turning in the same direction. There is a similar effect in the beam-rider mode.

Many years ago I had a model plane where this sort of track was the solution to a problem. The plane's rudder had half jammed, and it could not turn right. Worse still, it was also turning left gently in the absence of any command and was drifting in the wind in broad circles, rapidly receding toward the horizon. By not panicking (after the first minute or two) I managed a rescue by going for as long as I could in a gentle turn, then turning sharply left to complete 300 degrees or so of turn, then again flying a gentle left turn until the course was too far off, and so on. After a sweat-inducing fight against the wind and with a dwindling fuel supply, as well as the faulty rudder, the plane made it back to the home field. Mathematically inclined readers might like to estimate what forward speed can be maintained, given a certain flying speed and known radii of curvatures for left and hard-left turns.

Similarly, an occulting target (one that switches on and off, or is obscured periodically) can cause the Carpet Missile to steer along an unexpected track. Assume that the missile moves to the right when illuminated, left otherwise. With a periodically occulted homing target, the missile will begin to edge off to go on a curving track to the left of the straight-line track. Eventually the missile will be to the left of the beam and will then begin to simply move in a circle to the left, missing the beam that would turn it. An occulting heat source is used in infrared missile countermeasures (for example, on Air Force One, to prevent the president of the United States from being shot down by a terrorist infrared homing missile).

The effect of twin occulting targets is interesting. (This is also the principle of some missile countermeasure equipment.) The missile turns toward target A, then turns toward target B, in this way navigating to a position C that is neither A nor B (shades of the grand old Duke of York).* Alternatively, depending on the dwell time of the light from each source, the Carpet Missile may also go into paroxysms and chase its own tail.

*The nursery rhyme about the grand old duke refers to Frederick Augustus, Duke of York, an army commander famously unsuccessful in the wars with Revolutionary France in the 1790s: "Oh, the grand old Duke of York, he had ten thousand men, he marched them up to the top of the hill and he marched them down again. And when they were up they were up, and when they were down they were down, and when they were only halfway up they were neither up nor down." Maybe the duke was practicing missile countermeasures.

The simple layout given can be used in beam-rider mode too, simply by pointing the photocell backward and illuminating from the back with a wide beam pointing in the desired direction. There is a potential problem that the vehicle will deflect out of the beam, however.

More sophisticated guidance systems are typically needed to do beam-riding in practice, using two or more photocells. Even a vehicle with two photocells next to each other without further sophistication does not solve all the problems. For instance, when both are illuminated evenly, the machine progresses on a straight line, but with no illumination the missile simply stops, and it still slows down as it goes outside the beam. It may also weave from side to side badly just as the single-photocell vehicle does. The orientation of the plane of the receiver photo-diodes has an influence, as does the spacing between them, and the relation between motor speed and photocell illumination.

REFERENCES

Advisory Group for Aerospace Research and Development (AGARD). *History of German Guided Missiles Development.* Brunswick, Germany: Appelhans, 1957.

"Beamrider Missile Guidance Method." U.S. Patent #3782667. U.S. Patent and Trademark Office, Washington, D.C.

Churchill, Winston. *The Second World War.* Vol. 2. London: Cassell, 1949.

"Constant Bearing Course Homing Missile." U.S. Patent #3841585. U.S. Patent and Trademark Office, Washington, D.C.

Dunn, John H., Dean D. Doward, and K. B. Pendelton. "Tracking Radar." In Merrill Skolnik, *Radar Handbook.* New York: McGraw-Hill, 1970.

Gosling, William. *Helmsmen and Heroes.* London: Weidenfeld and Nicolson, 1994.

"Laser Beam Rider Guidance System." U.S. Patent #4111385. U.S. Patent and Trademark Office, Washington, D.C.

20 *Duohelicon*

Said the anti-clockwise Bindweed
To the clockwise Honeysuckle:
"We'd better start saving,
Many a mickle makes a muckle,
Then run away for a honeymoon
And hope that our luck'll
Take a turn for the better,"
Said the Bindweed to the Honeysuckle.

—Michael Flanders and
Donald Swan, "Misalliance"

People often call shapes like those in the accompanying diagram spirals, but they are of course helices or, if there is only one of them, a helix. They occur quite often in nature (as Michael Flanders and Donald Swann observe in the song "Misalliance" from their Broadway show *At the Drop of a Hat*), ranging in size from the tiniest virus to the largest tree, and the blueprint of life, DNA, is based on two intertwining helices.

In industry, spirals occur less often. Nevertheless, there are spiral drills, spiral worm gears, spiral conveyors, and spiral compressors (called screw compressors), and, of course, every nut and bolt in the world includes a spiral. But spirals don't seem to occur in vehicles much.

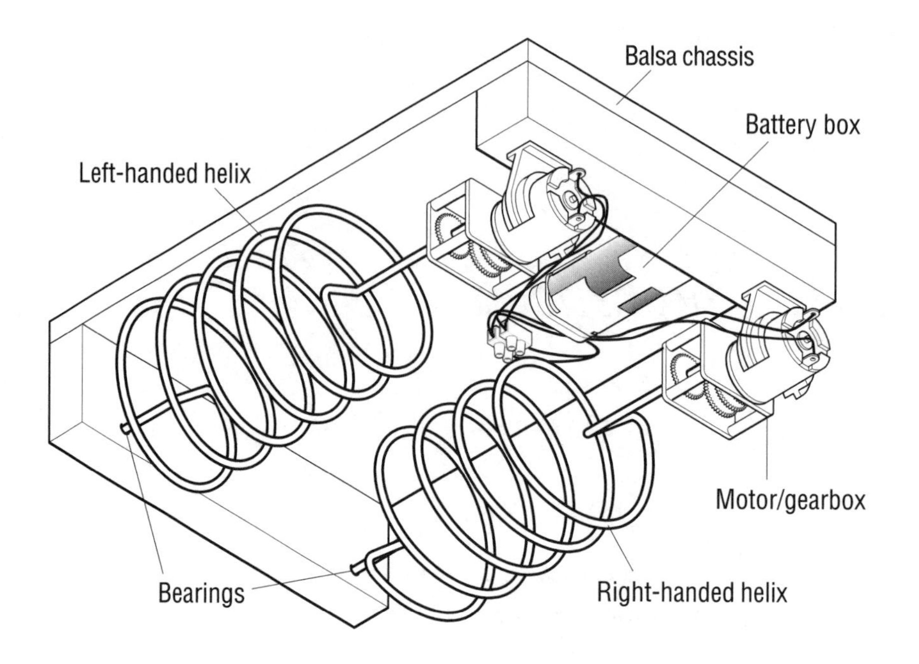

Pulp science-fiction writers have often depicted heroes such as Batman or Superman fighting deviously cunning villains who are equipped with a helical screw machine that looks something like a cross between a submarine and a large soil auger. In practice, however, such machines have not been possible, although there have been trials of machines that creep through the oozing mud on the bottom of oceans using this means of propulsion. Christopher D. D. Hickey's patented "Underwater Weapon System" has screws beneath it and looks a bit like the Duohelicon. However, Hickey intended that the screws would simply help his device bury itself deep in the mud and silt on the sea bottom, like a kind of mechanized flounder.

There are actually many real vehicles that propel themselves by means of a helix—ships and propeller-driven airplanes are the commonest examples. However, we do not recognize the helix in the propeller design because a propeller is only a very small fraction of a helix. A typical aircraft propeller with two blades, for example, is less than one-tenth of a turn of a helix, while even a ship's propeller is rarely more than about one-third of a helical turn. It was found early in screw-propeller experiments with ships that providing more than a half a turn or so did not increase thrust, while it did increase the power of engine needed, so multi-turn screws on Victorian ships were soon truncated to the designs we see today.

Our Duohelicon has helices with five turns on each, which partly explains why the vehicle looks so different. The other difference is due to the fact that the Duohelicon is a land vehicle. Because the inside of the helix would not contact the ground and would therefore do nothing, we have thrown it away, leaving the wire shapes you see.

The Degree of Difficulty

This is a more difficult project because of the fundamental need for complementary spirals made fairly accurately, one right-handed and one left-handed. You need some skill and strength to manipulate the tough springy wire successfully. You also need to be able to hook up the drive system and make it work smoothly.

What You Need

- ❑ 2 motors with transmissions
- ❑ Steel wire (piano wire or music wire)
- ❑ Plastic tubing to slide over transmission output shaft and steel wire
- ❑ Wood for body
- ❑ Alternative #1: Batteries, battery boxes (or a power supply like a car-battery charger running off the domestic AC supply)
- ❑ Alternative #2: Wire or radio control switch gear, as well as batteries and battery boxes (At its simplest, the wire control needs only two toggle changeover switches, a three- or four-wire cable, and a small box for mounting the switches. The switches should ideally be biased center-off toggle switches, the type that disconnect both outlets when centered and are biased to the center.)

What You Do / The Tricky Parts

This project does have tricky features. Don't forget that one helix must be right-handed, the other left-handed. Otherwise the Duohelicon will simply go around in circles. Each helix must be almost exactly constant in diameter, with no kinks or bumps (and if possible no big scratches). This is quite a feat. Probably the best way is to wind the wire around a circular drum mandrel of a little smaller diameter than you need for the helix. Wear gloves and goggles—if you inadvertently let go of the wire and it whips around it can be painful or even dangerous. Once you have wound the wire, you have the basic helix, and you can adjust how the ends of the helix finish off to connect to the drive system by bending them carefully with two pairs of pliers. (You will find it very difficult to correct faults in the helix itself with the pliers—any blemishes in the perfection of the helix tend to get worse if you try to correct them!) Using the same mandrel for both helices will ensure similar diameters. However, another tricky issue is achieving a good constant pitch (the distance between the turns of a helix), equal for each helix. Stretch the helix until a good pitch is achieved, but do it cautiously—you can't unstretch it.

The ends of the helix must be brought out pretty accurately along the axis. One end simply turns in a hole in the body or in a metal bracket, while the other is attached, perhaps via a short piece of stiff plastic tubing, to the transmission output shaft. The transmissions are fixed in place to the same chassis as the metal brackets. I used a piece of thin plywood as a base for the machine, with helices underneath, and put batteries above to give it some weight; however, many other layouts are possible.

The controls can simply be on/off on each helix, or they could be forward/reverse on each helix, or, most sophisticatedly, they could provide variable speed both forward and reverse on each helix. Rather than using trailing wires, obviously a radio control circuit could be used. These can often be obtained simply and inexpensively by taking apart a radio-controlled toy.

To test the corner turning of the Duohelicon you can simply slow one motor down (for instance, by removing a fully charged battery and substituting a more discharged one)—unless you have chosen to use a wire control or radio control system, in which case you can use the controls.

The Surprising Parts

On suitable carpet, you will be surprised how quickly the Duohelicon can propel itself in its curious way. However, try to go as fast as possible and you will soon discover that the Duohelicon has interesting alternative modes of behavior, including rapid sideways movement and a tendency to overturn. It certainly raises eyebrows as a vehicle too. More head-turning than a Ferrari, is my hunch, even though it is three orders of magnitude slower!

Using Your Duohelicon

The vehicle is happiest on a surface such as carpet, which has reasonably low friction and provides, as it dents slightly under the weight of the helix, an excellent grip.

The Duohelicon is at its worst on smooth surfaces, particularly sloping ones, where it will often slip out of control.

THE SCIENCE AND THE MATH

The Duohelicon performs well on carpet because the dents it forms amount to a sector of a machine nut, along which the machine can then progress smoothly (assuming its helices are reasonably exact) just like a machine screw into a nut. The need for a rather exact spiral follows because imperfections will lead the machine to spend much of its energy fighting itself. If, for example, there are radial bumps in the helices, then these bumps will form their own dents on the carpet, tending to shift the machine sideways instead of forward. In any case, the machine is rather inefficient, because it has to slide the helix underneath its own weight.

The Duohelicon tends to slip and go out of control on smooth surfaces because of the radical difference between static and moving friction. When an object is already sliding, the grip force between it and the surface it is sliding on is reduced markedly. In fact, if the helix is rotating rapidly enough, it will actually slide backward down a very gentle slope.

The Duohelicon will occasionally also crab sideways, particularly on hard surfaces, as the helix on one side wins a better frictional grip on the surface and pulls the machine over. It is not easy to control, except at the lowest speeds.

There is also the question of what happens when the Duohelicon turns. In fact the skidding action in turns is aided by the screw rotation. (Caterpillar tracks have no such assistance, and their awkward skidding action on hard surfaces makes track-laying vehicles difficult to drive on ordinary roads.)

The Duohelicon tends to run downhill very easily on hard surfaces. It is not difficult to see why: a simple cylinder rotating at right angles to the slope, if restrained from going sideways, will tend to slide down the smallest slope. This is because the frictional force F is rotated around so that it acts largely to prevent rotation, and only a small component of it still acts to restrain the downward descent. The simplest theory of friction F between surfaces says that the force between them is proportional to the force F_s applied squeezing them together, that is

$$F = \mu F_s,$$

where μ is a constant, the coefficient of friction.

But in the case of the vehicle on a slope, there is a force proportional to the product of mass M, gravity g. However, this force is not perpendicular to the surface of the slope but is aimed at an angle θ away from this, where θ is the angle of the slope to the horizontal. So

$$F_s = Mg \cos \theta$$

and

$$F = \mu Mg \cos \theta,$$

where μ is the coefficient of friction, M the mass, g the acceleration due to gravity, and θ the angle of the slope.

But with the cylinder rotating at peripheral velocity W, and the cylinder sliding down the slope at velocity V, the component of friction acting up the slope F_u is only

$$F_u = \mu Mg \cos \theta \tan (V/W).$$

There is a component acting at right angles to the slope F_s, where $F_s = \mu Mg \cos \theta \tan (W/V)$. With rapid rotation, W is large, and for small backward velocity V, F_u is much smaller than F.

Simple frictional theory differs from that just given, because the coefficient of friction μ_s between surfaces that are stationary relative to each other is larger than the normal coefficient μ. However, this fact does not affect the phenomenon described because the surfaces are all moving in this case. Also, the simple difference between μ_s and μ would not account for the effect seen.

A similar theory will also tell you that ordinary wheeled vehicles will run backward down smooth slopes, even if their wheels are running forward, once skidding is initiated. You will be surprised that it does not require a very steep slope to see this happen. Backward skidding can happen on surprisingly shallow slopes, a fact that many people don't appreciate until they have to drive an automobile up an icy road with a gradient.

I had the misfortune to have an auto accident in exactly these circumstances. I engaged reverse gear and began to reverse my vehicle up a slope. However, snow had reduced the coefficient of friction and thus allowed the vehicle to slide forward down the slope, even while the engine was driving it backward. My collision with the lamppost at the bottom of the slope might actually have been avoided had I engaged a forward gear and matched wheel speed with the sliding and then attempted to steer to one side of the post. If only my knowledge of frictional theory had overcome my instinct, I could have avoided the accident and a huge invoice from the body repair shop!

And Finally, for Advanced Users

I couldn't think of any way of making a vehicle like this with just one helix. The design would be a lot simpler with just one, but I could not see any way that it would not simply turn itself over and over and go sideways. Perhaps some cunning arrangement of wheels or skis would do the job?

The Duohelicon is perfectly designed to go sideways if only one helix is employed or if two are functioning with a slight imbalance. The undesirable sideways tendency can often vastly exceed the intended forward movement! I tested out different spirals for the Duohelicon by using a single spiral mounted on a frame that could rotate around a central post, avoiding the need for making two spirals of any test type. On most surfaces the sideways thrust of the helix was considerable and always several times greater than the forward thrust, with the helix acting as a very efficient roller.

The use of a slippery coating on the spiral will help to minimize sideways forces, but perhaps at the expense of forward motion on hard surfaces! Behavior on hard surfaces is difficult to predict. There are large differences between metals, for example. Aluminum wire is often available as scrap after overhead power-line work has been carried out. Ask the workmen, and you will often be given a few meters of strands of this material, which is usually ⅛″ or ¼″ (3–6 mm) or so in diameter, fairly easy to bend and hence very suitable for helices. There is one problem: an aluminum helix does not work very well! This was not too much

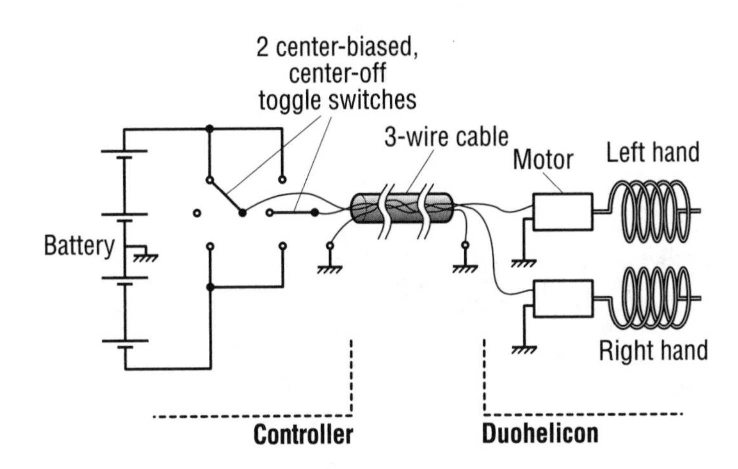

of a surprise to me. I once replaced the steel runners on my toboggan with aluminum strips to make it lighter, and because the aluminum would not rust. The result was a toboggan that was almost useless: aluminum surfaces offer much larger values of friction than steel surfaces do on ice. Cyclists will not be surprised either, as most bicycles with aluminum rims offer much more effective braking than those with iron rims. Carefully polished and varnished aluminum will work of course. But you probably won't want to try plain aluminum wire helices, unless to prove how bad they are!

I also found that on some surfaces, at least, it *does* matter which helix goes which way: reverse and forward are not equivalent. My version went happily in the forward direction but tended to shoot off sideways on some surfaces, and this tendency was much worse in reverse at high speed—try it and you may see what I mean. But why should it make any difference? Thinking about torque around an axis parallel with the helix axis may provide part of the picture. Think about a sideways wheelie . . .

The motion of the Duohelicon on uneven surfaces is a source of some fascination. If the vehicle ends up with one helix on a hard slippery surface, you might expect it to instantly squirm off course. But does it? You will have to build one to know!

REFERENCE

Hickey, Christopher D. D. "Underwater Weapon System." U.S. Patent #4566367. U.S. Patent and Trademark Office, Washington, D.C.

21 *Fishy Boat*

> If you stick a Babel fish in your ear, you can instantly understand anything said to you in any form of language. . . .
>
> Now it is such a bizarrely improbable coincidence that something so mind-bogglingly useful could have evolved by chance that some thinkers have chosen to see it as a final clinching proof of the non-existence of God. The argument goes something like this: "I refuse to prove that I exist," says God, "for proof denies faith, and without faith I am nothing." "But," says Man, "the Babel fish is a dead giveaway isn't it? It could not have evolved by chance. It proves you exist, and so therefore, by your own arguments, you don't. QED."
>
> —Douglas Adams, *The Hitchhiker's Guide to the Galaxy*

The Fishy Boat, is not, of course as mindbogglingly useful as the Babel fish that Douglas Adams describes. But the Babel fish is imaginary, whereas there is a Fishy Boat alive with tail kicking (at least when I switch it on) in my garage.

Today almost all powered boats are driven by screw propellers. So-called jet boats mostly amount to a screw inside a tunnel beneath the vessel, while older concepts like paddlewheels have all but disappeared. The so-called single-blade propeller used in some high-speed boats, which is a propeller operating only half immersed in the water, is a little closer to the Fishy Boat in concept. But only in oar-propelled boats do we see any departure from this propeller predominance. Closest to the Fishy Boat in concept might be the Venetian gondola, propelled by a large single oar at the rear, although in this case the more-or-less vertical oar is moved in the water in a more-or-less horizontal circle. The gondolier skillfully changes the angle of the blade cyclically so that the "lift" force of the oar is largely in a forward direction.

Fish of course usually swim by wiggling their tails from side to side. However, they operate wholly in the water, like submarines, whereas the Fishy Boat floats mostly above the surface, with only the tail fully immersed. Creatures that propel themselves while floating substantially above the surface are comparatively rare. Animals (including humans) that swim along the surface are also 95 percent underwater, leaving only the aquatic birds and a few insects in the true floating category, and none of these seems to employ a tail-wagging technique, preferring some variation on paddling.

The closest approach in nature to the Fishy Boat in concept might actually be flying fish. Flying fish can leap out of the water like dolphins and then glide on their greatly enlarged pectoral fins. However, they can also propel themselves in a continuous manner by skimming the water surface with their tails in the water and the rest of their body out of the water. They propel themselves by thrashing their tails, while their fins act as wings. The fish thus does not float but rather flies to stay out of the water.

The Fishy Boat might make passengers seasick were it ever made at full size. Each time the tail thrashes to the left, the boat thrashes to the left. (Perhaps the stabilized saloon and dining room installed by Isambard Kingdom Brunel on one of his early steamships could help passengers here. He had a stabilized saloon mounted on gigantic gimbals in the center of the ship. It was by all accounts technically successful at reducing seasickness among passengers, although probably at stupendous cost. The boat was abandoned soon after its construction, however, when it collided with the pier at the English Channel port it used.)

However, despite the disadvantage of its severe vibration, maybe the Fishy Boat could find a niche among vessels used on swampy inland waters, such as the Georgia-Florida Okefenokee Swamp. Certainly the model boat I made went well on a ponds and lakes with floating weeds—the fact that it had no propeller to get tangled up was a distinct advantage. Perhaps the fan-powered airboats that roar and splash their way noisily across Florida swamps could be replaced by a passenger-carrying silent wiggling boat.

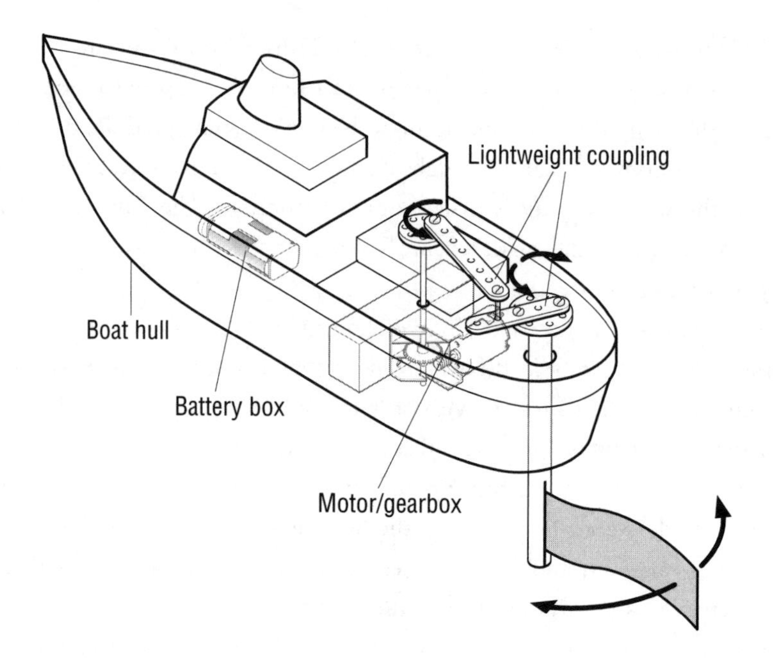

The Degree of Difficulty

This is a reasonably easy project, provided you feel confident about cutting up an old plastic boat with a razor-sharp utility knife and you have some dexterity with Meccano or Erector Set parts.

What You Need

- ❑ Plastic toy boat
- ❑ Toy motor with a moderate reduction gear (It should go several revs per second.)
- ❑ Battery and battery box to match motor
- ❑ Meccano or Erector Set parts

Various flexible tails to try, for which you might need:

- ❑ Semi-rigid plastic sheet (such as 0.5 mm polythene or polypropylene)
- ❑ Flexible rubber hose (such as LPG gas hose)

What You Do

Fit the motor and transmission unit with its smaller crank wheel and the larger crank wheel with its lever arm to the stern-mounted tail shaft. The connecting rod should be light in weight to allow the highest possible running speed.

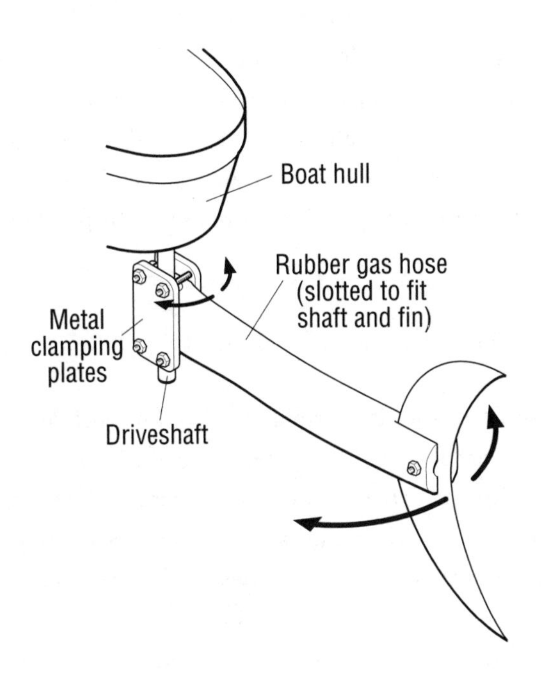

The Tricky Parts

The tail must be firmly clamped or glued or both to the oscillating shaft; otherwise it will slip. I used bolts to tighten it onto a sandpapered shaft, together with a very small drop of superglue.

The Surprising Parts

Why doesn't the boat just wiggle from side to side? Why does it go forward at all? And how is it that you can thrash the tail from side to side much faster than a fish and it still apparently works just as well?

Given that early steam engines were intrinsically reciprocating devices, I am a little surprised that early steamships were not propelled by means of a thrashing tail. Some of the earliest ships did indeed match the reciprocating steam engine with a reciprocating propulsive device, but generally this entailed some kind of paddling action, which is theoretically less efficient than a thrashing tail. The 1787 Delaware steamboat of pioneer John Fitch was a famous example of such a steam paddler.

Not only would the tail have been a good match for the back-and-forth motion of the steam engine, but it would have been far less cumbersome than huge paddlewheels and would have had far less drag on a ship that was using its sails. (Many early steamships spent most of their voyages sailing, using the engine only when becalmed in light winds or when the wind was from the wrong direction.) Perhaps ship designers were afraid that rhythmic twisting of the ship's hull would have led to failure of joints in the wooden planking and thus water leaks or worse.

Using Your Fishy Boat

Try longer and shorter tails. Try a rigid tail. Does it work as well as the flexible tail? Try a longer arm on the shaft (giving a smaller wiggle) or vice versa. The lower propulsive effort per stroke may be compensated by the greater speed possible from the motor. It is clear that some of the power supplied to the tail is wasted in producing waves: presumably these will be reduced if the tail is deeper, so the question arises, is a deeper tail useful?

You may be able to make the Fishy Boat go faster by using an airfoil type of tail, more like a shark's tail fins. The first part of the tail is thin and flexible, but the second part is much taller and more rigid. Even if it has the same area as the fishtail in the diagram, the taller fin may produce more thrust because is of its higher "aspect ratio," the ratio of length to width of an airfoil. This is an advantage in glider wings, because it minimizes the size of the tip vortices (which constitute wasted energy) relative to the airfoil area. This should be a good feature for a Fishy Boat, although whether you can show a real advantage will depend on a number of parameters—motor gearing, tail-beating frequency, length and weight of boat, size of tail, and so on.

THE SCIENCE AND THE MATH

The waggling tail, together with the hull, forms what is essentially a small part of a wave traveling in the bow-to-stern direction. A long flexible tail shows this clearly, and more clearly still if you have a stroboscopic disco light to illuminate it while it is thrashing back and forth.

The tail thrashes differently in the air than in the water. In the water the flexible tail is much more clearly forming traveling waves. The waggling tail can be thought of as behaving just like a screw propeller in two dimensions. If you project the shadow of a rotating screw onto a wall, it looks just like the waggling tail. (Scientists such as G. I. Taylor in the 1950s investigated the hydrodynamics of swimming bull spermatozoa, which, under a microscope, look just like rotating screws, because you see the sperm tails in silhouette.)

The behavior of the airfoil tail is slightly different: in this case, the larger, more rigid part of the tail is acting as a wing, while the flexible part is simply ensuring that the airfoil is swept through the water at the correct angle to provide forward propulsion.

Biologists who first investigated the speed of fish thought that fish showed roughly the same amount of drag that a rigid fish-shaped body would have. The steep increase in drag with speed means that small errors in the measurements would have made this a difficult thing to measure precisely. Later researchers looking at the amount of energy that fish used were surprised to measure fish using four times as much energy as they expected. What was happening? How had dozens of eminent scientists all been wrong? The explanation turned out to lie in some calculations done by James Lighthill in 1969. Fish do not behave like animated torpedoes. Because they don't have a propeller, they have to wiggle, and when they do this they disturb the "boundary layer" of water near them more than a stuffed fish would. The disturbed boundary layer makes them less efficient than a stuffed fish with a propeller.

It is difficult to play a ball game underwater because the ball is no good. A spherical ball slows very quickly. However, there is a stuffed-fish toy marketed for use in swimming pools that can be thrown remarkably far and fast underwater, showing just how fast a good streamlined shape can be with that undisturbed boundary layer. All you need to do now is figure out an efficient way of getting twenty-two scuba divers underwater at the same time!

When swimming fast, fish have Reynolds numbers of 50,000 and more, so water flows around them not smoothly but unsteadily, leaving a trail of vortex disturbances—what we call turbulent flow. The boundary layer is the layer of water near the fish, which is moving at a speed intermediate between the general body of water and the fish itself. You can think of the boundary layer as a kind of halo of water partly attached to the fish. Simply speaking, if the fish has a thin boundary layer, then the net size (fish + halo) will be smaller, and the fish will see less drag. Conversely, with a thick boundary layer, then the net size (fish + halo) is larger, and the fish will experience more drag. It turns out that the wiggling of the fish increases its boundary layer size, thus increasing its drag and increasing the amount of energy it needs to propel itself by a factor that is of the order of four for trout-sized fish. Hence biologists' anomalous results on the energy consumption of fishes were explained.

And Finally, for Advanced Users

It is clearly more work than changing the tail shape, but what about changing the hull shape? At the highest level, are longer hulls better than short hulls, and are deep hulls better than flat-bottomed shapes? Within these broad categories, what is the effect of loading in the hull? It is clear that loading the hull so that it is deeper in the water will slow the boat, but won't the tail be more efficient deeper and when operated from a boat that is heavier, with a larger moment of inertia?

And what about finer points of design, such as the exact planform, or the presence or absence of a keel fin? And what is gained by integrating the thrashing tail into the rear portion of a keel? If a bunch of Fishy Boats are available

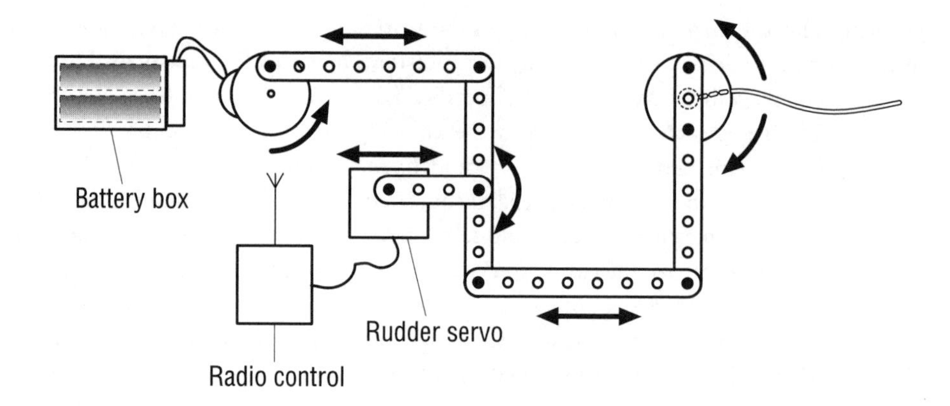

(preferably with the same motor, to be reasonably fair), you can have races to compare the performance of boats with different hulls and tails.

More easily perhaps, you could think about adding steering to the Fishy Boat via additional connecting arms, such that the tail continues to thrash from side to side but can also be biased to the left or to the right (see the diagram).

The tail can of course lie in the horizontal plane like the tail flukes of a whale rather than in the vertical plane suggested here. The advantage of this arrangement might be that a very high aspect ratio could be used, with long narrow flukes. But the efficiency may depend on ensuring that the tail does not come out of water, or even come close to coming out of the water. But how deep can you make the tail? And does it work better, the deeper it is?

Measurements on real fish and dolphins have revealed that fish such as salmon can travel at 10 m/s, with tuna able to sprint at 20 m/s. Even smaller fish such as sardines are capable of surprising speeds compared to their size—up to 30 body lengths per second. However, these fast fish move on the order of 0.5–0.7 body lengths per complete tail beat (back and forth). Smaller, slower fish use an approach more like our machine here, employing faster-beating small fins rather than a slower-beating tail. Many smaller fish also use their fins in a rowing action (forward with fin parallel to the streamline, backward with fin orthogonal to the streamline). This has been carefully measured and is less efficient than the airfoil tail approach, however, so is only suitable for low speeds.

REFERENCES

Adams, Douglas. *The Hitchhiker's Guide to the Galaxy.* London: Pan Books, 1979.

Alexander, R. McNeill. *Exploring Biomechanics.* New York: W. H. Freeman / Scientific American, 1992.

Robinson, D. *Animal Performance.* Open University Course S324. Milton Keynes, UK.

Videler, J. J. *Fish Swimming.* New York: Chapman and Hall, 1993.

Vogel, Steven. *Life in Moving Fluids.* 2d ed. Princeton, N.J.: Princeton University Press, 1994.

Webb, P. W., and D. Weihs, eds. *Fish Biomechanics.* New York: Praeger, 1980.

22 *Rotarudder*

Punting is not as easy as it looks. . . . He was evidently a novice at punting, and his performance was most interesting. You never knew what was going to happen when he put the pole in; he evidently did not know himself. Sometimes he shot upstream, and sometimes he shot downstream, and at other times he simply spun round.

—Jerome K. Jerome, *Three Men in a Boat (to Say Nothing of the Dog)*

Anyone who has tried propelling a boat by pushing on the riverbed with a pole will recognize him- or herself in the vignette in the chapter epigraph. The dominant mode of movement often seems to be rotation, and both the pole and boat seem to have wills of their own! Perhaps Jerome K. Jerome's three heroes would have fared better with a Rotarudder.

It is well known, particularly to golfers and table tennis players, that a spinning ball propelled through the air will not go in a straight line but will deflect to one side, as if steered by an invisible rudder. There is a similar but much stronger effect with a rotating cylinder. Such a cylinder can behave as an efficient airfoil. A rotating horizontal cylinder can produce "lift" in the same way that a wing and a rotating vertical cylinder can produce thrust from a transverse wind, as a sail does. This principle was used in a rather impractical full-size ship by German inventor Anton Flettner. Despite problems, the ship was a substantial vessel that was driven by wind pressure on its two tall rotating cylinder-sails, and it successfully crossed the Atlantic in 1925. There have been more recent attempts to use the principle, but none so far has been a commercial success. Here we demonstrate the principle to provide a simple and effective steering system for radio-controlled model boats.

The Rotarudder system represents what it is probably the simplest possible radio-controlled model boat. A toy boat is converted by the addition of a small rotating cylinder driven by the output of a single-channel on/off radio signal. When the radio signal is off, the cylinder is stationary and the boat's slightly skewed rudder turns it left. When the radio signal is on, the boat is driven to the right by the Magnus effect acting on the rotating cylinder. To maintain a straight-ahead course, the operator must pulse the transmitter on and off in a regular, rapidly repeating sequence.

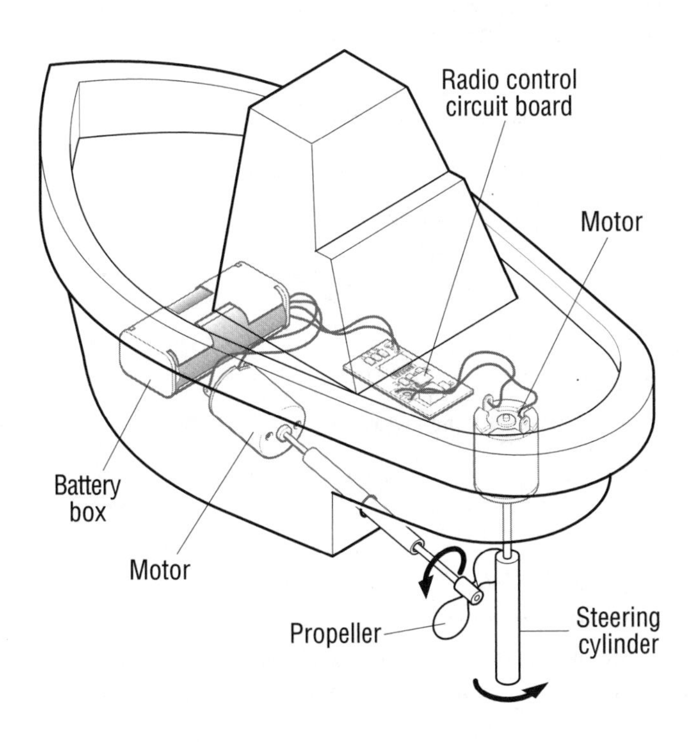

Alternatively, perhaps a simple automatic pulse generator, an electronic multivibrator, can be used for pulsing the motor. No servo-motor unit is required, no analog radio control link is needed—just an on/off switch on the transmitter and a receiver capable of responding to it. This kind of arrangement dates from the early thermionic-tube days of radio control, when it was picturesquely described as the Galloping Ghost system. Galloping Ghost systems were used to provide rudimentary control for model aircraft, controlling not just the rudder or ailerons, but the elevator too.

The Degree of Difficulty

This is a reasonably straightforward project, particularly if, as suggested, the radio control is cannibalized from an old toy, and manual pulsing of the transmitter is used to provide the straight-ahead setting.

What You Need

- ❑ Toy boat
- ❑ Toy single-channel radio control (A simple forward / turn backward toy car can be as little as $7 or $8 from Tandy or Radio Shack.)
- ❑ Diode
- ❑ Resistor (5 or 10 ohms, to run the motor slowly)
- ❑ Small cylinder

For the automatic straight-ahead pulsing system:

- ❑ 3 transistors
- ❑ 6 resistors
- ❑ 2 capacitors
- ❑ Variable resistor
- ❑ Center-off center-biased toggle switch
- ❑ Circuit board

What You Do

To cannibalize the radio-controlled toy, simply remove the automobile electronics connecting the output to the motor via a diode (this ensures that the Magnus cylinder motor will not reverse). Be sure that when you press the transmitter on-

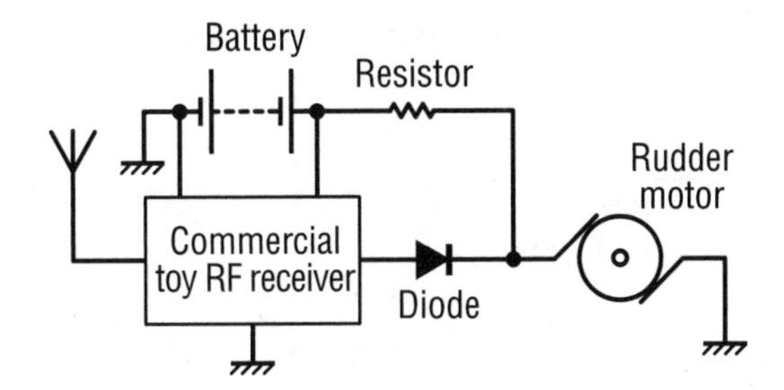

switch the motor simply speeds up without changing direction. The resistor in the circuit connected directly to the battery allows the motor to run slowly even when the receiver is not activating it. Choose the resistor value to be just sufficient that the motor runs very slowly—a few ohms should be about right for a small motor. A piece of resistance (nichrome) wire could be used instead of a resistor.

Once all looks well, place the boat in the water and check that the skewed rudder gives a steady, permanent, not-too-sharp left turn. Then check that switching the transmitter on continuously gives a reasonably sharp right turn. Then launch the boat, pulsing the transmitter on and off, increasing the amount of on time if the turn is too much left and vice versa. (See the advanced-user section for alternative ways of arranging the control.)

The Tricky Parts

These little toy motors typically don't like reversing direction or starting and stopping—it is better if the motor rotates all the time in the same direction, simply running faster or slower. The apparently obvious alternative circuit, in which the cylinder rotates alternately clockwise or anticlockwise, is thus not generally desirable. You can achieve this by including a resistor as indicated, which allows the motor to keep rotating slowly even when the radio receiver output has reversed (and the diode is blocking reverse current flow).

The easiest way to operate the rotating-cylinder rudder is through a hole in the bottom of the boat. The snag is that the boat will sink if you make a hole in the bottom, obviously—unless you plug it up again. I plugged up a large (6 mm) hole by actually gluing the motor in the hole to block it. I found that only a little water will actually get in via the motor bearing. A purist could place the motor at the top of a vertical shaft that reaches above the water line.

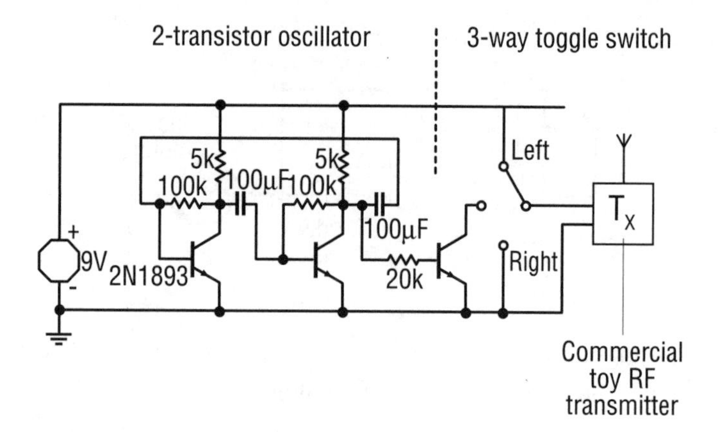

The Surprising Parts

The boat turns quite sharply when the Magnus cylinder is rotated, even with a rather weak motor using only 3 volts or so. I found I could easily turn the boat in its own length.

Using Your Rotarudder Boat

Children find pulsing the transmitter difficult. Constructing the pulse unit to do this automatically is obviously useful. The pulser here (see diagram) gives two or three pulses per second, at which frequency the speeding and slowing of the motor can be heard distinctly. The switch simply changes from T_X off (left turn) to T_X pulsed (center) to T_X on (right turn). The variable resistor adjusts the on/off ratio for a particular rudder setting, so that the boat goes straight ahead.

The single propeller has a tendency to give the boat a fixed left turn anyway (with a right-hand screw on the propeller), which is another effect that needs to be taken into account. In many cases, this avoids the need for the use of an auxiliary left-turn rudder bias.

The use of a smaller cylinder, or of a flywheel, will ensure that the motor runs more steadily, even when switched on and off rather slowly.

THE SCIENCE AND THE MATH

The Rotarudder works because of the Magnus effect: a rotating cylinder produces a deflection of an oncoming airstream. At zero rotation speed, the cylinder provides simple drag. As the rotation speed increases, the cylinder deflects the incoming airstream more and more, until at very high speed the airstream is deflected through about a right angle.

One way of thinking about why this happens is to remember the Bernoulli equation—see the Bernoulli's Clock project—which says that when the velocity of a fluid is increased, the pressure decreases, and vice versa. Think of fluid moving from left to right past a cylinder rotating clockwise. The rotating cylinder drags the fluid forward at the top, increasing its speed, while at the bottom it pushes the fluid forward to some extent, decreasing its speed. Following Bernoulli, we can then see that a lower pressure will be created at the top of the cylinder and a higher pressure below, these pressure zones deflecting the fluid stream above and below the cylinder downward.

The Magnus effect is described in aerodynamics texts like that by Douglas et al. I estimate that the "lift" force L (component of force at right angles to the boat axis) produced should be approximately given by the following formula:

$$L = 2\pi R^2 \omega \rho V_0 h,$$

where R is the cylinder's radius, ω the angular speed of the cylinder (radians/sec), ρ the fluid density, V_0 the boat speed, and h the height of the cylinder, which is assumed to be large compared to the radius. This equation also assumes both zero fluid viscosity and zero turbulence, a somewhat unreal pair of assumptions. To include the effects of viscosity and turbulence makes the whole problem intractably difficult, however.

The Magnus sideways "lift" effect from the rotating cylinder is thus increased a lot at high values of R. But the drag of the rotating rudder is also increased at high values of R. Increasing the rotation speed of the cylinder does not increase drag on the boat but will require more rudder motor power. Interestingly, the sideways force increases with forward speed of the boat, so that, like a normal rudder, the Rotarudder does not lose its power at high speed.

There are additional subtleties. The wash from the propeller of the boat design here streams directly over the cylinder, increasing its effect. Using tiny pieces of artificial snow in the water, I could see almost a 90-degree turn in parts of the backwash with full speed on the cylinder. You can use plastic chips of roughly neutral buoyancy from a water-filled snowstorm or glitter toy if you want to try this yourself.

The electronic pulser is a classic oscillator "flip-flop" or multivibrator circuit. The two symmetrical transistors and their components are the oscillator. There are two states: transistor 1 on/transistor 2 off, and transistor 1 off/transistor 2 on. Each state is stable for only a short period while the capacitor charges up; then the transistor that was formerly off is switched on, which switches off the formerly on transistor. The third transistor acts to pulse the transmitter circuit.

And Finally, for Advanced Users

If you don't like the idea of the pulsed on/off radio system, you could cannibalize a slightly more sophisticated radio-controlled toy in which the transmitter and receiver system is capable of communicating not just on/off, but on/off/on in reverse.

You can try the same effects with a bigger diameter cylinder (but beware—although the motor power requirements are small, they do grow with the size of

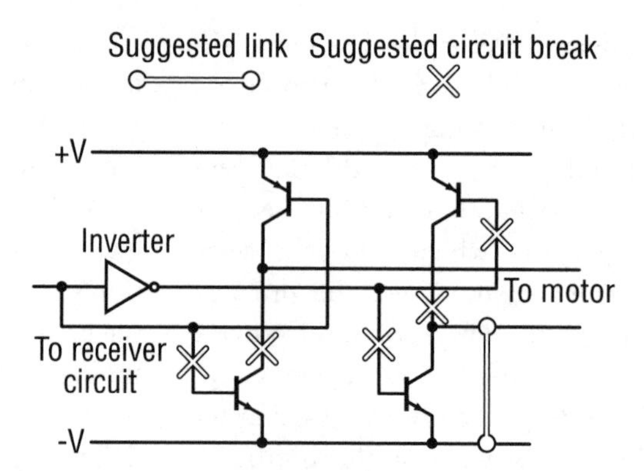

the cylinder). You can also make the cylinder longer, but this may cause problems with the stability of the rotation and could even cause capsizing in a higher wind.

A more subtle variation on the cylinder design is to add sandpaper, or small slots or flaps. Do these simply increase the effective diameter of the cylinder, or do they behave in a different way?

You could build a much bigger Rotarudder than this—perhaps big enough to steer a passenger-carrying motor boat? It would be ideal if the engine has some sort of electric generator on it, as inboard marine motors and more sophisticated outboard motors do, as then the power needed for the electric motor required to rotate the rudder cylinder would not be a problem. You might be able to find cheap 12 V motors of a few tens of watts power at a junkyard. If a high-power rudder is needed, boat stores sell trolling motors for electric outboard propulsion (used for quiet low-speed maneuvering when stalking fish).

If you are used to dealing with transistor circuits, you may be able to improve in several ways on the arrangements described. For example, if you can identify parts of the drive circuit of the radio-controlled toy, you may be able to avoid the use of the diode and gain some useful voltage for the motor operation without adding extra batteries. Generally, the output circuit of these toys is a four-transistor push-pull output. You can disconnect two of these and short out one of them (see diagram).

The on/off pulser could be used at higher frequencies, so that variations in the motor speed due to the pulsing are imperceptible. This is the principle of "pulse-width" DC motor controls commonly used in industry. There may be slow-acting components in the circuit of the toy, such as power supply decoupling capacitors, which will need to be removed or reduced in size for this approach to work.

REFERENCES

Douglas, J. F., J. M. Gasiorek, and J. A. Swaffield. *Fluid Mechanics.* London: Pitman, 1979.

Jerome, Jerome K. *Three Men in a Boat (to Say Nothing of the Dog).* Baltimore: Penguin Books, 1976.

23 *Cable Yacht*

For I dipt into the future, far as the human eye could see,
Saw the Vision of the world, and all the wonder that
would be;
Saw the heavens fill with commerce, argosies of
magic sails.

—Alfred, Lord Tennyson, "Locksley Hall"

Perhaps Lord Tennyson had seen the stack of aircraft circling the runway at Chicago's O'Hare airport through a time warp and in "Locksley Hall" was describing flotillas of jumbo jets. But maybe he had in mind Cable Yachts, a more realistic technical prospect when he was writing, more than sixty years before the Wright brothers.*

Californians have also been in the vanguard of applying technology to reduce impact on the environment. One or two power stations in California have been having some trouble recently finding electricity to send along power cables. Maybe they can instead utilize the cables for an ecologically sound transport system of the future—the Cable Yacht.

*Tennyson goes on more darkly to presage air combat and aerial bombing in the twentieth century: "Heard the heavens fill with shouting, and there rain'd a ghastly dew / From the nations' airy navies grappling in the central blue."

The Degree of Difficulty

This is an easy project, requiring only a small number of Erector Set parts, and a team of two people to handle both ends of the cable. The only real point of difficulty I can foresee concerns the weather—you do need a moderate breeze.

What You Need

- ❑ Clothesline or its equivalent
- ❑ Pulley wheels
- ❑ Pulley shafts
- ❑ Erector or Meccano Set parts or wood for other parts
- ❑ Spring washers
- ❑ Polystyrene sheet
- ❑ Modeling clay or a bag of stones for a payload

What You Do

First, build the four-pulley "bicycle" as indicated. The pulleys should run smoothly, with just a millimeter or two of clearance between them and the cable, with the cable running between them. The polystyrene sheet airfoils can be shaped with a large sharp smooth-edged kitchen knife, unless you happen to possess a hot-wire cutter. The airfoils are taped to an Erector Set strip, which is mounted on a bolt with spring washers not fully tightened down. This allows it to be rotated, albeit with a very stiff action, to set it at an angle to the wind.

The system works in this way. The Cable Yacht is attached to the cable, and the weight (or a package for delivery by your yacht) is hung from the bottom, as in the diagram. The cable is then stretched between two operators, ideally with the cable roughly at right angles to the breeze. The operators should be holding the cable up at a comfortable height, but reasonably high. The airfoil is then set at about 30 degrees or so to the wind, and the yacht is given a gentle push. It will accelerate away from the pusher. When it reaches the other operator, the stiff swivel is used to reverse the airfoil, and the yacht can be returned.

> **CAUTION**
>
> When you are holding the end of the cable and the Cable Yacht is heading toward you, be prepared for the impact. Wear gloves if you have built a large device. You will also find it a useful trick to slacken off the cable just as the yacht reaches about 2–3 m away, so that the increase in slope slows the yacht down.

The Tricky Parts

There are no tricky parts, except some care to ensure that the pulleys are arranged close together so that the yacht cannot run off its cable.

The Surprising Parts

The small sail provides the package with a surprising amount of speed—20 mph (35 kph) is easy in a moderate breeze, appreciably faster than the wind itself

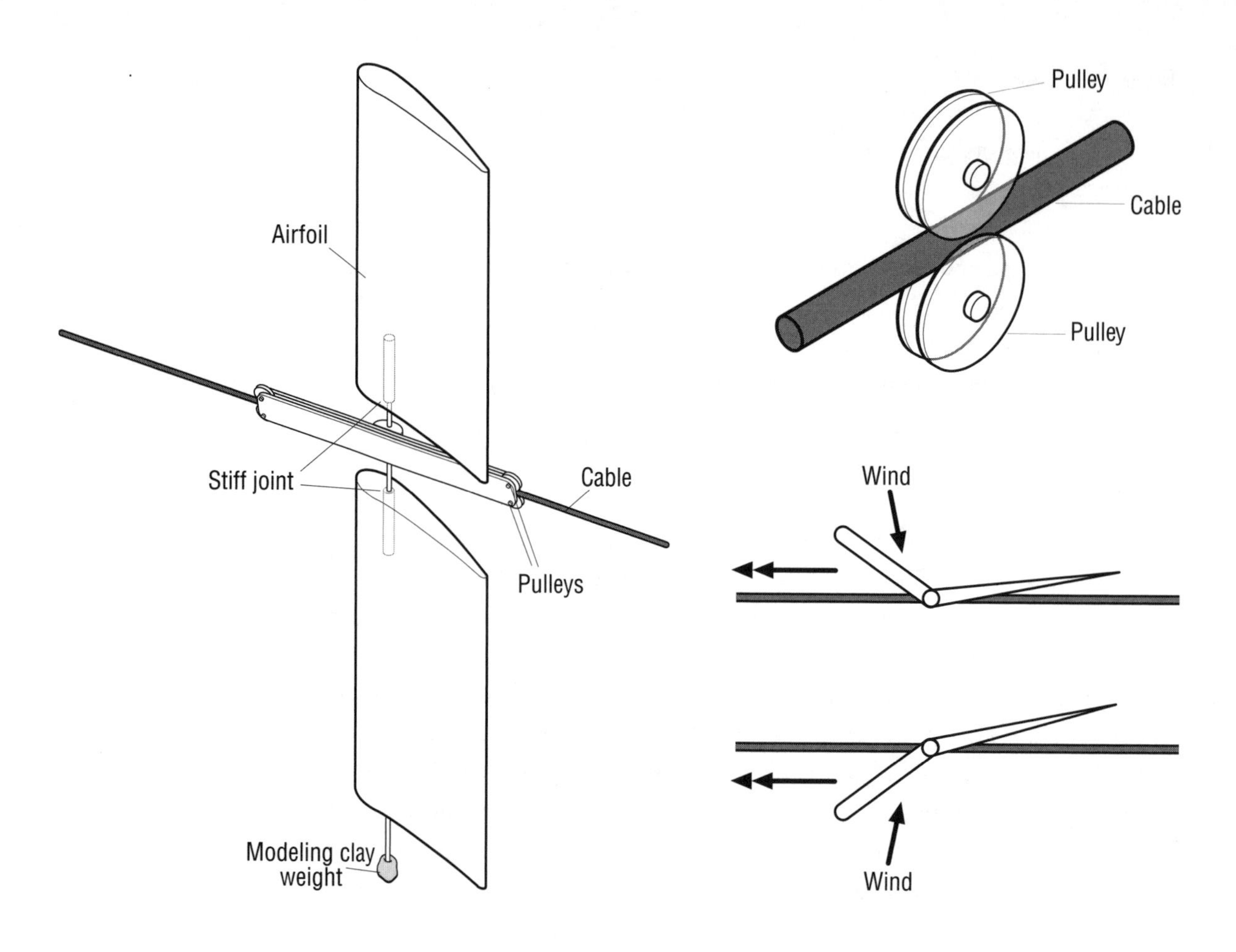

going cross-wind, and often faster than you can run. Many people imagine that sailing vessels are incapable of high speeds because they have only the wind to power them. In fact, the limitations on their speed are almost all to do with the fact that they sail on water. Sand yachts and ice yachts give a true indication of the high speeds that are possible using sails. From early beginnings in the Netherlands and Baltic Sea in the 1760s, ice yachts spread to the United States. The Hudson River hosted races in which, by the 1900s, sailing at three or four times the wind speed, ice yachts exceeded 100 mph. And the *Guinness Book of Records* in 1938 listed the fastest official speed for an ice yacht, "Debutante," as 143 mph! Like the ice yacht, the Cable Yacht has no water to contend with, and it is also capable of surprisingly high speeds.

Even in quite a stiff breeze, the Cable Yacht can be weighted so that it is stable. On the basis of tests indoors with a powerful electric fan, I had expected the Cable Yacht to roll badly and had even intended to equip it (with enormous increases in complexity) with two cables. This was not necessary, as a single-cable device works very well.

Using the Cable Yacht

A chain suspended between two points on the Earth will lie in a catenary curve. Similarly, the suspension cable of our device lies in an approximate catenary curve if the yacht is fairly light in weight. If the yacht is fairly heavy, then the cable will take the form of a shallow V-shape, while the Cable Yacht itself will follow a curve that is in theory an ellipse. These curves, although different in detail, both mean that the yacht will run downhill to start with, which is handy for starting it off, and will run uphill at the end, which is handy for slowing it down. You can try exaggerating the curve by slackening the cable to see its effects.

The same principle is used in many subway systems. The cars run on solid rails screwed to rock in the ground, of course, rather than on a cable. The rails, however, slope downward into a dip between stations, just like a cable, to provide a natural acceleration and deceleration action.

You will find that different airfoil angles are needed to work with the cable at different angles to the wind. How close to the wind can you sail the yacht? You will probably find that closer than 30 degrees or so is impossible, and even at this angle, the yacht needs a firm push to get it going.

THE SCIENCE AND THE MATH

The Cable Yacht works in an obvious way that needs little basic explanation. However, questions such as how fast it can go and how much weight it can carry are interesting and not so obvious. Yachts of all types go at their highest speed not when running downwind, which is what you might naively expect, but rather at approximately 90 degrees to the wind. A wind on the beam when sailing is sometimes known picturesquely as a "soldier's wind," because it provides fast and easy sailing, even for a novice sailor.

Suppose the wind speed is V. Running downwind, a yacht can go at best only as fast as the wind, so that $V_{max} = V$. But running with a 60-degree angle to the wind, if the air were a relatively solid body, we might expect the yacht to move at a speed of $V_{max} = V/\cos 60 = 2V$ (think of the airfoil being pushed aside as it slices through a moving slab of butter).

The cable lies in a catenary curve before the yacht is added. Approximating this to a section of a parabola, we have

$$Y = K(X^2 - XL + 0.25L^2),$$

where Y is the downward deflection of the cable, K a constant proportional to w/T, X the distance along the cable, supported at $X = \pm L/2$, w the weight per unit length, and T the tension in the cable. The maximum droop occurs where $X = 0$, when $Y = 0.25KL^2$. The droop thus grows rapidly with the square of cable length, limiting the unsupported length.

If the cable were very light, and the Cable Yacht very heavy, the cable would lie in a V-shaped valley with the yacht at the bottom when it was added. The slope angle of the V would be given by

$$\sin \alpha = W/2T,$$

where W is the weight of the yacht. For example, with a yacht of weight 1 kg force and a tension of 20 kg force, $\alpha = 1.4$ degrees. Again the cable length is limited, although the droop in this case grows proportionally only to length rather than to length squared. With a 1 m maximum droop in the example, the cable could be up to 80 m long unsupported. The yacht in this case follows an elliptical path between the two points of support. You can quickly show this as follows: push two thumbtacks into a piece of paper, tie a piece of string loosely between them, and, holding the string taut with the tip of a pencil, slide the pencil along the string. Or you could show the elliptical form via a simple simulation using a computer spreadsheet, if you want a more accurate check.

Clearly the actual shape of the cable and the actual slope encountered by the Cable Yacht lie somewhere in between these two extremes. In both cases, however, it is clear that a higher tension will bring benefits in terms of smaller slopes and less droop in the cable.

The efficiency of the airfoil is another tricky question. The airfoil needs to be a symmetrical section in order to work just as well in either direction along the cable. It also needs to work at a large range of angles to the wind, in general. This makes its choice problematical. A "variable geometry" airfoil may well be the most efficient, for example, one that can be modified to give a concave side toward the wind on either side. A flexible textile sail automatically achieves this, although sails often have difficulties with the turbulence around the leading edge. Some sailing yachts try to solve this by using the mast as part of the leading edge of the sail, employing a teardrop-section mast and a sail that wraps around the mast or fits in a slot.

It is easier to choose a planform for the airfoil. In general, long thin planform, "high aspect ratio" airfoils (narrow chord, long length) will tend to give faster speeds than short fat ones. The drag on an airfoil is proportional to the width of its chord, whereas the lift (thrust in this case) is proportional to its area. Another way of thinking about this is to consider the wing tips as creating vortices (and thus drag) proportional to their chord (width parallel to flow), while the wing area creates the lift: long, thin planforms have more area per unit tip width.

And Finally, for Advanced Users

As discussed earlier, the basic Cable Yacht will go faster if you tension the string more. More impressive increases in speed can be obtained by sailing with the wing closer to right angles to the wind direction on the ground. The closer the airfoil or sail approaches right angles, the faster the airfoil can travel relative to the wind speed. It is a little like pushing a frictionless nut along a threaded shaft that is confined from axial movement but free to rotate. If you put more threads

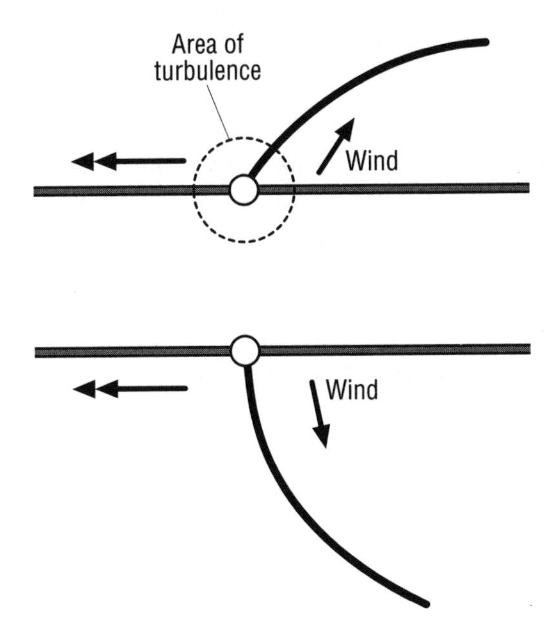

per inch on the nut and shaft, and push forward on the nut the same distance as before, the bolt unwinds further.

In the limit, perhaps surprisingly, you can actually sail even with the wing exactly at right angles to the wind direction on the ground. This is actually what happens inside the Wells turbine. The Wells turbine has symmetrical airfoils that are placed exactly at right angles to the airflow. Whichever way the air flows, once the Wells turbine is rotating, it continues to rotate in the same direction, most efficiently the direction in which the blunt end of the airfoil points, even when the airflow reverses. (These turbines are used in generating power from ocean waves, by using the column air pushed in and out of a cylinder whose lower half is placed in the ocean.) Why does the Wells turbine work? The key to this conundrum is that the airflow in the Wells turbine is axial along the tube, but, relative to the moving turbine airfoils, it is still moving in a direction to urge the airfoil forward.

With two cables, the machine might be slightly better stabilized, which might also increase performance. But you would need to get the tension and spacing of the two cables just right, and I decided that it was too difficult to manage easily when sailing the Cable Yacht on a cable held by hand at each end.

For long distance use, the Cable Yacht clearly needs to be able to pass support masts. This immediately creates two difficulties: getting up enough speed to be able to go over the hump of the mast and, once there, supporting the string in such a way that the pulleys will go over it.

The latter problem has to be addressed by all cable-supported or cable-powered vehicles, from aerial cable cars and ski lifts to trolleys. A typical solution is to attach the string to the edge of a thin metal blade, cutting away the pulley support frame so that the blade can pass through one side of it.

REFERENCE

Ross, David. *Power from the Waves.* Oxford: Oxford University Press, 1995.

Antediluvian Electronics

24 *Beard Amplifier*

Marconi has produced a new electric eye.

—Sir W. H. Preece

Sir W. H. Preece, engineer-in-chief of Britain's Telegraph General Post Office, was exhibiting Guglielmo Marconi's improved coherer detector for radio waves at the Royal Institution, London, on June 4, 1897. The phrase he used, "electric eye," is redolent of the magic of radio in those early days. The new detectors were lauded as giving humankind a sixth sense as fundamental as sight, sound, smell, taste, or hearing.

Amplifiers are quite fundamental to life. Consider the act of writing I am now engaged in. It is an example of mind over matter. My thoughts—minute electrical signals in my brain—on their own would be too weak to raise the hind leg of a gnat. But my brain and body amplify them using a cascade of neurons. Each neuron activates several more neurons, which each in turn activates several more, and so on, and the effect is to increase the size of the electrical signal before it reaches sets of muscle cells. Once there, muscles convert the electrical signal into mechanical contraction force, and that force moves my finger. My invisible, microscopic, and weak thought energy has been amplified until it is large enough to affect the macroscopic visible world.

Amplifiers are of course used in almost every technical device today, but this is a rather recent phenomenon. Before the 1950s there were no transistors, and before the 1920s not even vacuum tubes, and these are the basis of all electronic amplifiers. It is a little-known fact that E. Branly, Marconi, and other pioneers of radio communication used amplifiers in their radio receivers as long ago as 1890. They were very important in those days, and J. A. Fleming, in *Electric Wave Telegraphy and Telephony,* describes them in some detail (pages 468–493). These amplifiers, called coherers, were inefficient, to be sure, but nevertheless offered thousandfold or millionfold increases in sensitivity compared to the spark gaps used by physicists such as Heinrich Herz when radio waves were first discovered.

Coherers, sometimes called Branly coherers, were tiny glass vials with contacts inside connected to the antenna wires. An incoming radio wave causes brass or platinum filings inside the glass vial to cohere or stick together slightly, sufficient to cause a battery connected via the vial to operate a bell or Morse buzzer or printer. The vast majority of the energy—a joule or so—needed to operate the Morse device came from the battery. Only a few microjoules or less of radio energy input were needed: the system was an amplifier. Instead of being limited to a few hundred meters range, for the first time long ranges could be achieved. Within a decade, radio waves would span the world, thanks to the coherer.

This project, a "magnetic coherer" or Beard Amplifier, was inspired by these early radio amplifiers. The "beard" of iron filings that hangs from a magnet or electromagnet when it is swept past a surface contaminated with filings is a common observation. (On the workbench in my garage, with a rather untidy mixture of pieces of broken equipment and a sprinkling of sawdust and filings, it is an especially common observation!) It seems unlikely that no one has previously discovered the relay or "transistor" action that we are trying here, particularly as there was so much work on this at the end of the nineteenth century. However, I certainly have not seen an account of it. Perhaps its limitations are such that, when it was discovered in the past, it was quickly forgotten. Nevertheless, the device certainly operates, and it illustrates many of the difficulties of amplifying devices.

The diagram shows the simple principles of the device. A variable current on the input circuit varies the magnetic field in the two coils and hence the degree of solidity of the beard of iron filings in between the two magnet poles. The output circuit applies a voltage between the pole pieces, and an output current that depends on the input current is thereby obtained. The device functions because the filings form hundreds of current paths between the two pole pieces, which grow in number and decrease in resistance as the magnetic field intensifies. When the magnetic field is reduced, these paths are substantially broken as the filings mostly fall off the upper pole piece under gravity, or at least they are less strongly pushed together.

The Degree of Difficulty

While not for the complete beginner, the reader with only modest engineering talent and little equipment should find this straightforward.

What You Need

- ❑ 2 large iron bolts (for example, 5″ long and 10 mm diameter)
- ❑ 2 solenoid coils to fit over the bolts*
- ❑ Plastic collar to fit over the bolts
- ❑ Controlled current supply for the input (I used a variable number of 1.2 V rechargeable batteries.)
- ❑ 12 V DC electric supply for the output
- ❑ Lamp for the output
- ❑ Iron filings (but see the advanced-users section for alternatives)

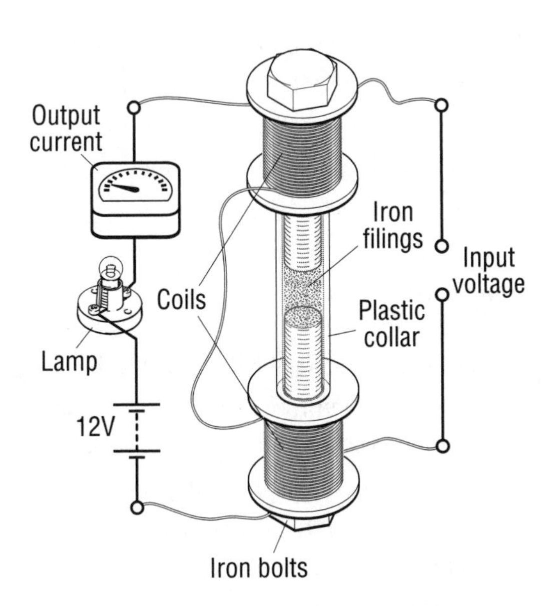

*These coils can be wound from a long piece of varnish-insulated wire—about 0.08 mm diameter enameled copper wire #40 AWG or #44 SWG, 150 m or 200 m long, for example. Alternatively, take apart old small transformers to use the already-wound bobbin or wire, as I did. They should be of at least 500 ohms resistance each.

What You Do

Two electromagnets wired in series form the input circuit, the direction of the connection being such that a north pole is formed on one and a south pole on the other. A variable resistor or other variable power supply is connected to the input. The electromagnets are placed close together, with a small volume partly filled with iron filings between the poles and an output circuit arranged with sufficient battery voltage that more current will flow in the output circuit than flowed in the input circuit or, at least, that more power (current × voltage) will be present in the output than was in the input. The output circuit can be just an ammeter, or a 12 V lamp, perhaps with an ammeter as well in series.

The Tricky Parts

You need to adjust the spacing between the two electromagnets carefully, so that (a) there is enough space for gravity to pull the filings back down on the bottom pole, but (b) there is the smallest possible distance for the filings to have to climb to reach the upper pole. Only when correctly adjusted in this way will the Beard Amplifier show positive power amplification.

The Surprising Parts

When you apply the weak control current, which gives rise to a very weak magnetic field (it will barely pick up the lightest paper clip, for example), a current nevertheless passes through the controlled circuit. With correct adjustment, the output circuit controlled could be fifty times as high in power as the input, in my set-up.

Using Your Beard Amplifier

Marconi's coherers could perhaps better be described as bistables than amplifiers. They sat in a nonconducting state until a large enough radio signal came along, then changed to the conducting state, which persisted until the glass vial containing the filings was jogged to shake the filings apart again. This was often achieved automatically by arranging for the coherer to shake itself. The vial was jogged by the bell or buzzer it was operating. Some of the same bistable effects also seem to occur in the Beard Amplifier.

To start using the Beard Amplifier initially, you may need to tap the filings gently to arrange them into a suitably loose heap. Also, you may find it useful to adjust the distance between the pole faces. Too far apart, and little output current will flow; too close together, and there will be a strong tendency for the output current to remain on in the absence of input.

THE SCIENCE AND THE MATH

When the axial magnetic field is applied to the iron filings, they have a tendency to reorient themselves along the field lines. This can easily be understood if it is remembered that a magnetic field will induce magnetic poles on an iron rod or other magnetic material placed in it. In our case, each filing becomes a little magnet, in effect. Assuming the filings have a more-or-less elongated shape, they will form a north pole at one end and a south pole at the other. These induced poles will then be attracted by the applied field, like the tiny magnetic compasses they are, to lie in neat lines pointing in the direction of the applied field.

They also tend to join together in long lines between the poles. Again, induced magnetic poles are the answer to why this happens. The induced poles attract each other, with the closest filings first joining, and then these larger clumps joining up in turn. (If you sprinkle some filings on the surface of some viscous pancake syrup and then watch them move when you put a magnet close to them, you get a better idea of what happens inside the Beard Amplifier, although there it happens much faster.)

Once the filings have formed a bridge, it is obvious how an amplifying action can occur. The greater the magnetic field applied, the lower the resistance of each bridge. The harder the filings press together, the lower the resistance of the circuit they form, since the contact area increases as the force between them increases. The total resistance will be dominated by a few high-resistance links. Suppose, for simplicity, that there is just one resistive contact area, area A, length L, whose area is proportional to the force applied (as is exactly the case, for example, for a pneumatic tire pressing on the ground):

$$A = kF.$$

But F is proportional to the magnetic field B applied, so

$$A = k'B,$$

where F is the force between the two filings, k, k' constants, and B the magnetic field.

Then resistance R is given by

$$R = rL/A = rLk'/B,$$

where r is the resistivity of the iron filings. There is an output current $I_{out} = V/R$, so

$$I_{out} = Vk'B/(rL).$$

And finally, since the magnetic field B is proportional to the input current I_{in} with constant, say K, we have the amplifier formula

$$I_{out} = I_{in} \, VKk'L/(rL).$$

Now in fact many simultaneous parallel current paths form, where the first chain of filings forms a bridge that conducts, say, 1 microamp, the second conducting too, giving 2 microamps, and so on, with the output current proportional to the effective number of conductive links. Or, in other words,

$$\text{Output current } I = I_b N_{eff},$$

where I_b is the current up an average bridge, and N_{eff} the effective number of bridges. If the number of links were very large, in the region of millions or more, there would be a strong averaging process. Unfortunately, the number of links is smaller than this, and this fact leads to a source of noise in the amplification process, as the output current tends to change in jumps as new bridges of filings form and fall apart.

The device can only be an amplifier if the variation in the secondary circuit power (current × voltage) exceeds the variation in power in the primary current. The current that a set of iron filings will bear is limited. If too much current is forced through the filings, they will oxidize or even ignite, and the resulting iron oxide is substantially an insulator.

Also limited is the solidity of the beard of filings that a weak electromagnet due to a small input current will form. I used two coils taken from transformers that took only a few mA of current at a volt or two, giving an input circuit power of a few milliwatts. The output circuit I used operated on 10 or 12 V and worked at a minimum resistance value of around 100 ohms, giving a maximum output power of about 1 W. My gain factor varied over orders of magnitude depending on the delicate setting of the pole gaps and also varied randomly in time to a large extent. Nevertheless the gain factor often exceeded ten times and went up to a maximum of fifty times or so with favorable adjustment.

There is also the question of the linearity of the device: is the output current accurately proportional to input, or does it simply flip on and off like a switch? The latter is certainly not the case: the device output current certainly varies with input current, and in my measurements it certainly increased steadily. The question of linearity of response is very difficult to answer on account of the noise and hysteresis that seem to be intrinsic to the device operation. Hysteresis is the electronic equivalent of backlash in gears. Depending on whether you make a measurement going up or going down in input, the output current equivalent to a particular input current will be different. Hysteresis could arise in a number of ways in the Beard Amplifier, perhaps from the magnetic remanence of the filings (the effect that allows permanent magnets to exist—the leaving of permanent magnetic poles in material after an electrically imposed field is removed). Another mechanism could be microwelding. Particularly in sparsely formed beards of filings with a high current flowing, microscopic welds could form in the filing bridges, which will only be broken apart at much lower input magnetic fields than they formed at (or when the filings are jiggled).

And Finally, for Advanced Users

The Beard Amplifier could be modified in a number of ways. Rather than a jiggling action provided only when a signal pulse is received, as used by Marconi and his contemporaries, perhaps a continuous jiggling, either from a mechanical device or perhaps from a superimposed AC magnetic field, could be tried to avoid bistable action.

Also fairly obvious is the use of a permanent magnet in part of the device. If arranged in the correct sense, a Beard Amplifier can be made to display an inverting amplification as well as a noninverting amplification.

The "noise" in the Beard Amplifier output circuit is, quite simply, very bad. Superimposed on the amplified input signal are random fluctuations that, depending on the exact nature of the set-up of the device, can easily exceed the wanted signal in magnitude. We tend to forget today that modern transistor and integrated circuit amplifiers also add noise to a signal. Careful design is still sometimes needed to avoid problems from this noise—for example, in the input amplifiers of radio and TV receiver equipment. Perhaps the use of large pole piece areas and smaller pole gaps might yield a lower noise figure, by averaging the noise due to the forming and breaking of electrical paths through more filings. (My tests in this direction were inconclusive.)

The question of how fast the Beard Amplifier will function was also not clear from my tests. Although Marconi's coherers were designed to receive radio signals at hundreds of kHz carrier frequency, they could not carry information—Morse code—at more than a few pulses per second. Similarly, I could not persuade my devices to function at more than a few Hz. Audio and even electric-power-line frequencies seemed beyond them. However, the application of more ingenuity here could perhaps pay dividends: the use of an HF magnetic bias (magnetic tape recorders use such a scheme to improve recording quality) or the use of more finely divided filings might work, perhaps in an inert or even reducing atmosphere to obviate oxidation.

Another area where experimentation might bear fruit is in the nature of the filings used. Some early coherers used brass filings, but clearly these would not work here, being nonmagnetic. Iron with plating that resists oxidation may help—or the use of filings of another magnetic metal such as nickel that resists oxidation. (Fairly pure nickel is readily available in Canada in the form of Canadian coins, by the way—most other countries use nonmagnetic alloys of copper, or copper-faced iron.) The use of a liquid might be possible with nickel, which would resist electrolytic corrosion to some extent if water or an alcohol were used. The use of oil is probably possible—iron filings in thin oil will still conduct electric current to some extent—although there may be problems because of the tendency of the filings to become coated with the insulating oil.

Of course the shape and size of the filings will also have an influence. I used filings with irregular shapes from a grindstone. But these were curly shapes and little spirals often formed when a metal is filed. (Look at filings of different materials using different tools under a low-power, 30× microscope if you want to know the sort of thing I mean.)

REFERENCES

Branly, E. *Comptes Rendus*, Vol. 111, 1890, p. 785.

Fleming, J. A. *Electric Wave Telegraphy and Telephony.* London: Longmans, Green, 1916.

Marconi, G. "An Improved Coherer Detector for Radio Waves." UK Patent #12,039, 1896.

25 *Tornado Transistor*

They that sow the wind shall reap the whirlwind.

—Hos. 8:7

The weather—its storms, its clouds, rain, and snow—is an amplifier, a principle the ancient Jewish tribes of Israel had obviously grasped, based on the comment quoted in the chapter epigraph. Apocryphally, a butterfly flapping its wings somewhere in Hawaii will determine whether a massive hurricane will wipe out the banana crop in Central America or damage vacation homes in Florida a few weeks later. This amplifier effect is the reason for the notorious inaccuracy of weather forecasting. Famously, in 1987, Michael Fish of the British Meteorological Office announced to England that there was definitely no chance of a serious storm. A day later a million trees were uprooted and traffic disrupted in southern England for two days while roads were cleared. Fish had clearly not taken into account that butterfly no. 2,389,123 on Ascension Island had fluttered its gossamer wings four times instead of three times!

In the strange new version of the transistor described here, the input current—of electricity rather than wind—stirs up a little storm in a teacup, which then provides an output current.

The ordinary (silicon) transistor receives an input current and controls an output current. As normally used, the current going into the base-emitter junction controls a current flowing between the collector and the emitter junctions. The transistor is almost the simplest imaginable amplifier and is, of course, useful not

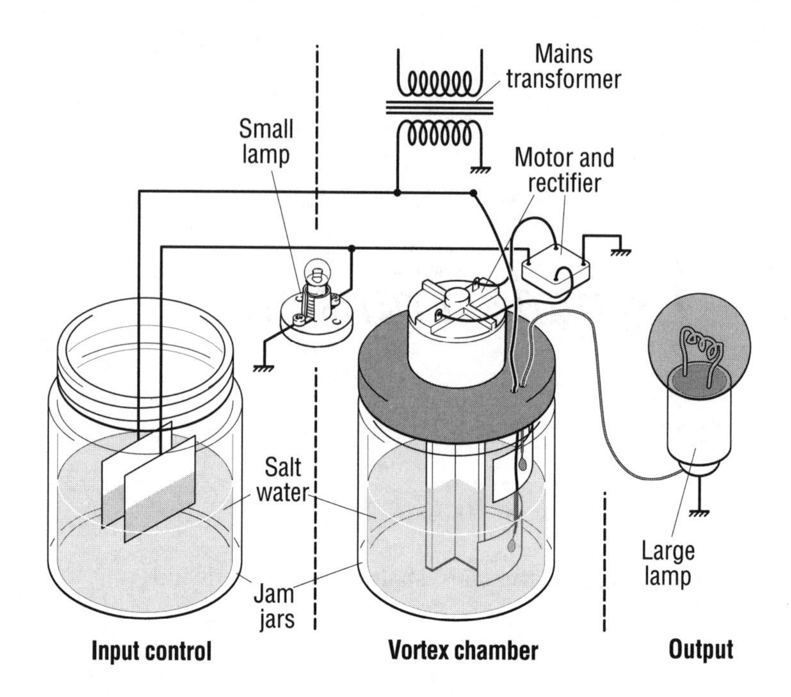

just practically, but also for demonstrating the principles of amplifiers generally, for learning purposes.

The problem is that all this action in the silicon transistor takes place in a few microns of silicon—you can't actually see anything. In the nineteenth century the electronic amplifier was the relay, and you could hear the buzzing and clicking and see the contacts moving (T. E. Herbert and W. S. Procter, *Telephony*, 2:85–169). Even in the 1920s, with vacuum tube amplifiers (Herbert and Procter, *Telephony,* 1:864–899), there was something to see. When the amplifier was on, the filament in the thermionic tubes glowed, and you could make out the different parts and imagine the electrons inside flowing between the electrodes. But with silicon, all the magic is completely invisible.

This unfortunate fact was the reason I devised the Tornado Transistor. How the Tornado Transistor input works is plain for all to see, while the Tornado Transistor output current is conveniently at a level large enough to light a large lamp or turn another motor (or even the input to another Tornado Transistor).

The Degree of Difficulty

This is an easy project, at least by comparison to a do-it-yourself silicon transistor. No delicate manipulation of minute whiskers of metal or ultrapure semiconductor is required. Nor do you need a billion-dollar silicon chip wafer–fabricating factory! However, you do need to be able to marshal and connect some low-voltage electrical equipment, and you need a household AC transformer, so the project is not for the absolute beginner.

What You Need

- ❑ Small low-current electric motor (I used a 12 V type, but electric motors sold for use with solar panels are also okay.)
- ❑ Bridge rectifier (Small 1 A or 2 A types rescued from old electronic power supplies should be fine.)
- ❑ 110 V (or 240 V) → 12 V electric power transformer (A 12 V, 1 A type, perhaps rescued from equipment, should be fine.)*
- ❑ 12 V vehicle lamp or 12 V electric motor (for an output load)
- ❑ Balsa wood for stirrer paddles
- ❑ Pieces of tin plate (such as from a cookie can)
- ❑ Wires and alligator clips (or use solder)
- ❑ 2 glass jars
- ❑ Salt
- ❑ Water

What You Do

Set up as suggested in the diagram. The input to the electric motor is controlled by a saltwater controller in a glass jar. This is simply two plates of metal or carbon rods dipping into an aqueous common salt (sodium chloride) or copper sulfate solution. As you dip them in deeper, more current will flow. You could of course use a commercial variable resistor in place of this; however, beware, you will need a low-resistance, high-wattage variable resistor, like those in the controllers used for model slot-car racing. In any case, surely the saltwater controller is more appropriate, given the use of saltwater in the vortex transistor itself.

The motor, when operating, stirs the water so that it forms a vortex surface rather than its usual flat one. This vortex surface, to a greater or lesser extent, covers the gap between the two output plates, providing a current path for the AC power supplied to them, which is thus (roughly) proportional to the input current.

*An automobile-battery charger is fine, but you have to connect to the 12 V AC connection inside, not to the external 12 V leads that go to the battery normally.

CAUTION

Be sure that you don't allow salty water near the 110 V/240 V electric power input. Do as I did and connect the transformer close to the power outlet, and take a low-voltage lead off a reasonable distance to the rest of the gear.

The Tricky Parts

You do need to get the amount of salt in the water about right, as well as (more obviously) the level of water. With too little salt, the output circuit will simply be starved of current and the current gain will be too low. With too much salt, the transistor will tend to switch the output circuit from completely off to completely on rather than provide an analogue of the input.

The Surprising Parts

It looks horrible after a few days: a jar of swirling, gurgling water with rusting connection plates. But it works! And, like a real transistor, it actually amplifies, in this case on the order of twenty times.

Using Your Tornado Transistor

The Tornado Transistor can be used in exactly the same way as its micrometer-sized little silicon brothers. For a graphic demonstration of the amplifying action, power a tiny 12 V instrument lamp with the input current, perhaps in parallel with the vortex motor, then connect the Tornado Transistor to control the larger output current. You could use the Tornado Transistor to operate a sys-

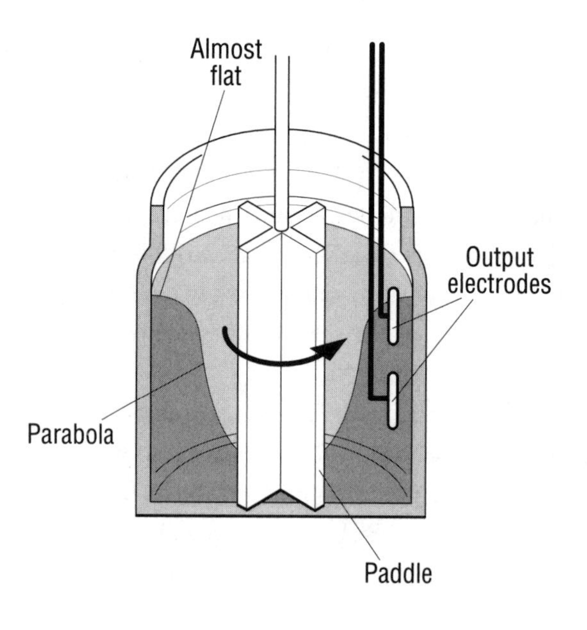

tem where a small input (say, from a solar cell) operates a large output (say, a motor to draw a pair of curtains).

You can do many of the tricks that a semiconductor transistor allows. You can put two of them in series to improve the gain. Maybe you can connect the output around back to the input via a delay (add a flywheel to the motor?) to make an oscillator.

THE SCIENCE AND THE MATH

A silicon (bipolar) transistor has three connections: the emitter, the base, and the collector. As normally used, the base is used for input signals, while the emitter is the common connection (often ground) and the collector is used both to feed in power via a resistor and as the output, as the diagram shows.

A semiconductor transistor can be thought of as a current amplifier: the emitter-base current (in forward bias) is multiplied by the gain Hfe to give the collector-emitter current. In the Tornado Transistor, the current to the stirring motor corresponds to the emitter-base current, while the output current is that through the salty water in the vortex. There are other points of similarity with the silicon transistor: the liquid level at the edge of the jar can be thought of as the voltage on the base—the higher it is, the more current will flow in the output circuit.

The system I put together for most of my demonstrations had a current "gain" of at least twenty times or so, when run on 12 V AC. It would operate an automobile lamp of about 5 W (500 mA output current) using about 25 mA of input current to the stirring motor.

An amplifier does not create energy of course. The energy in semiconductors comes from the DC voltage supply, and here it comes from the AC domestic supply.

You can get more power out of the Tornado Transistor by simply applying more voltage to the output circuit. In theaters in the past, saltwater baths were used to control the brightness of theater lighting. Tanks with many liters of salty water were used, and they could control lamps of up to a kilowatt or so capacity. (Behind the curtains, it was pretty hot and sweaty in the days when I controlled our school theater lighting with 4-ft-long variable resistors. But imagine the scene backstage in the good old days, with all that heat, and salty water gently steaming as well! These days, for theater lighting we have triac controllers, which are ten or twenty times more efficient and thus don't get hot with the wasted power.)

Of course there are limits to the extent to which this can be carried out. In the semiconductor transistor, the voltage at which the semiconducting silicon breaks down is the limit. In the Tornado Transistor, too high an applied voltage would eventually boil the salty water. With a silicon transistor, applying too many volts is a more disastrous problem: a huge current will flow, melting the unfortunate device and destroying it almost instantaneously.

Some interesting mathematics arise in the shape of the surface of the vortex. In a rotating bowl, the surface assumes a parabolic shape. The easiest way to understand this is to look at the force F on an imaginary particle of water on the surface. This force F must be at right angles to the surface, otherwise that particular particle of water would move toward the position of least energy—flying outward or inward and increasing the surface level X until E was a constant.

$F = M\omega^2 R \cos\theta$ (centrifugal force component along slope)

$= Mg \sin\theta$ (gravity force component along slope),

where M is the mass of the particle of water, R, g the acceleration due to gravity, X the depth of the water, ω the angular velocity, and dX/dR the slope of the surface. $\sin\theta/\cos\theta = \tan\theta = dX/dR$.

So $M\omega^2 R = MgdX/dR$,

and so $MgdX = M\omega^2 R\, dR$

and $dX = (\omega^2/g)R\, dR$.

Integrating, $X = \frac{1}{2}(\omega^2/g)R^2$.

This simple picture is disturbed in a vortex such as we have here by the fact that the jar is not rotating (this would make construction and the electrical connections more difficult), but only the central stirring paddle. Water near the jar walls is actually stationary, leading to a vortex shape that flattens out near the wall, as the diagram indicates.

And Finally, for Advanced Users

It is of course perfectly possible to make smaller Tornado Transistors—the smallest ordinary motors I could find are those made for activating camera auto-focus and are just 8 mm in diameter, although there are smaller "stepping motors" used in watches, just 3 mm across, and maybe these could be made to work. Baby Tornado Transistors made with these transistors would operate faster but would of course have a lower output current capability, just like their semiconductor transistor cousins.

Further analysis ought to be able to predict the input-output characteristic of the Tornado Transistor. Noting the power required to stir at different speeds will give the rotation angular velocity as a function of input current. This can be translated into the depth of a paraboloid formed on the surface of the brine. By integrating with respect to radius, the volume of the paraboloid-surfaced cylinder of water can be calculated and hence the depth to which the output electrodes will be submerged for each input current. Knowing then the conductivity of the brine and the output-circuit applied voltage and load resistance, and making assumptions about the current pattern, the current flowing in the output circuit versus depth of immersion can be calculated. Finally, all these calculations can be combined to give the overall characteristic. Whew!

REFERENCE

Herbert, T. E., and W. S. Procter. *Telephony.* 2 vols. London: Pitman, 1934.

Electric Water

26 *Meltdown Alarm*

Hot ice and wondrous strange snow.

—William Shakespeare, *A Midsummer Night's Dream*

You can help avoid the occasional but irritating failures of your freezer by using what the folks at T-Fal Cookware, in their advertisements, dub the "appliance of science."

The meltdown in your freezer can be used to generate its own alarm signal in a very simple way: by measuring the electrical resistance of a frozen salt solution as that solution melts. The enormous change of electrical resistance of a salt solution when it freezes seems to be an obvious sensor, but one that few have thought about, and one that seems to be unused in industry.

An ice cube frozen from slightly salty water can easily show a few kilohms to many megohms transition over a very small temperature range, allowing the use of the simplest of circuits. A couple of small wires can be led out from a food freezer, for example, without disturbing the magnetic rubber air-sealing strip too much, to sound an alarm if the contents become too warm.

The Degree of Difficulty

This is a smorgasbord project: you can make it anything between almost trivially easy (the ice-cube and multimeter alarm) and quite subtle (the position-sensitive meltdown alarm).

What You Need

For the simplest possible gadget:

- ❑ Multimeter
- ❑ 2 wires
- ❑ Ice cube mold
- ❑ Salt

For something a little more sophisticated:

- ❑ 2-transistor oscillator (2 transistors, 4 resistors, and 2 capacitors—see the diagram)
- ❑ 2 electrodes
- ❑ 2 wires
- ❑ Piezoelectric sounder (the kind that needs an audio-frequency AC drive)

What You Do

The simplest possible set-up is a multimeter with a beeper wired to slightly salty water frozen in an ice cube. As the temperature rises, the cube will melt in a matter of twenty minutes or so, and the resistance measured will fall from hundreds of thousands of ohms to below 1,000 ohms, giving an unequivocal meltdown signal.

A low-current buzzer or light-emitting diode (LED), for example, could be powered directly in exactly the same set-up, using a battery instead of the multimeter's internal battery, and larger metal plates soldered to the wires to give a somewhat larger current. If the normal state is frozen, then there will be electrolysis and corrosion in the cell only when the alarm signal is on, and this will be modest at low currents.

Electrolysis can be avoided by using a low voltage, below 0.5 V. Alternatively, avoid corrosion by using carbon rods as electrodes (from used cheap non-alkaline batteries). You can get such a low voltage by simply connecting a diode in series with a single-cell battery, since the diode, if of the silicon type, will typically drop the voltage by 0.6 V. A single nickel cadmium rechargeable cell will give out 1.2 V, which will be reduced to 0.6 V after a diode.

The ultimate solution to the corrosion or electrolysis problem is to avoid it entirely by using AC current, however. It can also be avoided by using a piezoelectric sounder driven by a separate oscillator, isolating the ice cube from direct

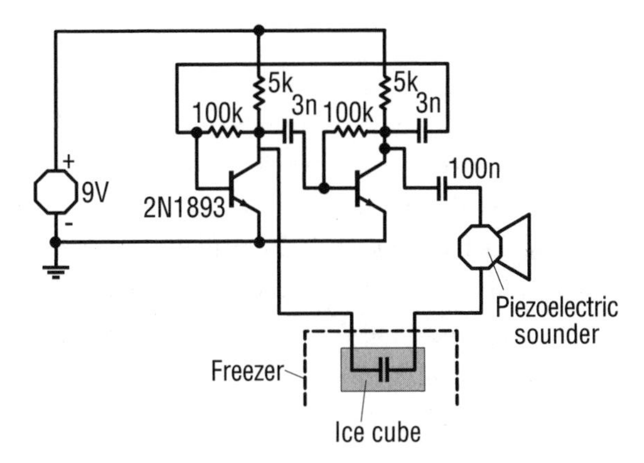

current by using a capacitor, which still allows the AC current that is needed to operate the sounder to go through, of course.

For specific purposes—for example, for sounding an alert when your food freezer is too warm—the freezing point can be adjusted by changing the amount of salt added (see "The Science and the Math" section), although this also changes the liquid ("on") resistance.

There are other parameters in the meltdown alarm that could be varied to advantage. The common salt, sodium chloride, could be changed for others. The solvent could include small amounts of antifreeze, for example, to lower the freezing point while not reducing the "on" resistance too much.

The range over which freezing occurs will be affected by the nature of the liquid as well as its additives: for precise operation a narrow freezing range is desirable, while for a "thermometric" use a wider freezing range would clearly be an advantage. Some substances have peculiar behaviors during melting: for example, one sample of ice cream I tested actually increased in resistivity during melting!

The Tricky Parts

There is little to go wrong in this demonstration. However, it is worth bearing in mind that, at room temperature, an ice cube forms a film of conductive liquid water on its surface as soon as it is taken from the freezer. Even though 99 percent of the cube is still frozen, the alarm warning that melting point has been reached will still go off. So an alarm will not show its "off" state unless the sensor ice cube is placed in the freezer. This effect makes the sensor more sensitive and quicker to react when the freezer door is left open, for example.

The Short Explanation

For a substance to conduct electricity, there must be electrically charged objects that are free to move in an electric field. Ice is a solid and, like most solids, consists of rigid arrays of molecules, none of which can move. When an electric field is applied, nothing basically happens; ice is an insulator. A solid metal, of course, behaves quite differently, because although the metal is a solid with all its atoms locked together in the same way, the outer electrons of each atom of the metal are free to move. When these electrons move, carrying electric charges as they go, they constitute an electric current; thus metals are electrical conductors.

Water molecules in water and water-based solutions split (to a small extent) into OH^- and H^+ ions, which are charged and are free to move. Hence, when the ice melts, it becomes a conductor.

When salt (NaCl) is added to water, its crystals dissolve, releasing Na^+ and Cl^- ions, which are also charged and free to move in the water. Thus when ionic compounds such as salt are added to water, its conductivity is further enhanced, since there are many more charged ions to carry current.

THE SCIENCE AND THE MATH

The conductivity of ions in dilute solutions is fairly straightforward: the more conductive ions you add, the greater the conductivity of the solution. If the salt added to the solution is fully ionized, you can calculate the conductivity from tables that give the conductivity factor (conductivity/concentration) of each ion in the salt, adding them up. So sodium chloride conductivity of 126 S (siemens)* cm^2/mol is composed of 50 S cm^2/mol sodium ions and 76 S cm^2/mol chloride ions. However, when the solution is frozen, conductivity sinks rapidly to zero.

The conductivity of solutions of ions in water does have a number of curious quirks. First, the freezing process is not uniform and sudden but occurs in stages, via the formation of needlelike crystals that become more and more numerous until they fill all the bulk of what was formerly liquid. During the early stages of this process, conductivity can occur between the needlelike crystals, between volumes of liquid that are not yet divided by crystals. There is a further complication because the first crystals formed in brine are of ice containing less salt than is in the bulk solution (or even in fairly pure water), and this leaves the remaining solution a more concentrated, and thus more conductive, brine.

Of course, you don't have to use common salt or a similar fully ionized ionic compound. If the compound used is one that dissolves but does not fully

*The siemens unit is simply the inverse of the ohm: a conductor of 1 siemens will conduct a current of 1 amp when 1 volt is applied. It is named after Werner Siemens, founder of the giant German electrical firm and a pioneer of electric generators.

ionize, such as an organic acid like acetic acid HAc, then the conductivity factor is high in very dilute solutions, but lower in more concentrated solutions. This can be thought of as arising from consideration of the equilibrium of the ionization and ion neutralization reactions:

$$\text{HAc} \quad + H_2O \rightleftharpoons \quad H^+ \quad + \quad Ac^-$$

Acetic acid + water $\rightleftharpoons$ Hydrogen + acetate ion

The rate of the forward reaction will be proportional to the concentrations of HAc and H_2O, the rate of the backward to that of H^+ and Ac^-. If there is more HAc, then there will be more H^+, since at equilibrium,

$$R_f = k[\text{HAc}]\,[H_20] = R_b = k'[H^+]\,[Ac^-],$$

where R_f and R_b are the forward and backward reaction rates, k, k' are constants, and the square brackets indicate concentrations.

At low concentrations of HAc, essentially all of it splits up—the back reaction is very weak. However, as the concentration of HAc builds up, the back reaction builds up and not all the added HAc ionizes, so the conductivity factor (which assumes 100 percent splitting into ions) shrinks at higher concentrations. Water is itself only weakly ionized and has the surprisingly low conductivity of 4.3×10^{-6} Sm^{-1} (or a resistivity of over 20 megohm cm) when it is absolutely pure.

The freezing point of a solution is lower than that of the pure solvent because the molecules of the solvent like to join up in exact, pure crystals (some substances are purified in this way). The "foreign" molecules get in the way and slow down the formation of those crystals. Now, the formation of crystals is a dynamic equilibrium between formation and dissolution. Crystals are both dissolving and forming at the same rate at equilibrium (at the freezing point). Anything, such as the foreign molecules, that interferes with the formation process but doesn't interfere with the dissolution process (and, since the foreign molecules are not incorporated in the crystals, they don't) results in a lower melting point. The standard formula for estimating approximate effects on freezing point is

$$T_{mo} - T_m = X\mathfrak{R}T_m^2/H_f,$$

where T_{mo} is the normal melting point, T_m the lowered melting point, X the gram molecular concentration of the additive, $\mathfrak{R}$ the gas constant (8.31 JK^{-1}), and H_f the latent heat of fusion of the solvent. The alert chemistry-minded reader will have spotted that this formula, because it includes X, offers a way of estimating the molecular weight of an unknown compound: by adding the known mass of the compounds M, then measuring $T_{mo} - T_m$, X can be calculated, and

$$\text{Molecular weight} = M/X.$$

And Finally, Advanced Meltdown Alarms

A high-temperature alarm—a fire alarm, for example—could easily be arranged by employing the change of resistance in a solid electrolyte. Sodium chloride in a tube is also a possibility, although it would not be very sensitive, as NaCl melts only at 800°C. Maybe a low-melting salt such as sodium sulfate decahydrate, with a melting point of 32°C, would be a better choice.

Position-Sensitive Meltdown Alarms

With a technique reminiscent of the method used by electrical engineers to find faults in cable insulation, you can also make a position-sensitive detector—a hot-spot detector for freezers, for example, or a fire-location detector. Essentially, the idea is to use the melted spot as the "wiper" in a potentiometer. A potentiometer is simply a resistance with, in addition to the two end connections, a wiper connection that can slide along to any point between the two ends. The two ends of a resistance are connected to an input voltage, and an output voltage is taken from one of those ends and the wiper. The output is always a fraction of the input voltage, the fractional factor being determined by the wiper position. The wire of the potentiometer here is a simple piece of thin nichrome wire. It is placed down the middle of the long trough or cylinder of frozen, slightly salty water, along with a parallel piece of copper wire. Alternatively, if you use a steel trough, you can dispense with the copper wire, but you need to ensure that the nichrome is correctly supported continuously along its length. You could use a piece of sponge to support the nichrome below the surface of the water before freezing.

The battery is connected in series with a resistor to provide just a few hundred millivolts of "sensing" current. (Any more than this, and the sensing current will melt the ice anyway!) On melting the water in the trough at any point, the conducting water connects the nichrome wire to the copper or the metal trough.

The voltage required along the nichrome to read out the position of the meltdown need only be 100 mV or so, obviating any serious electrolysis effects. The

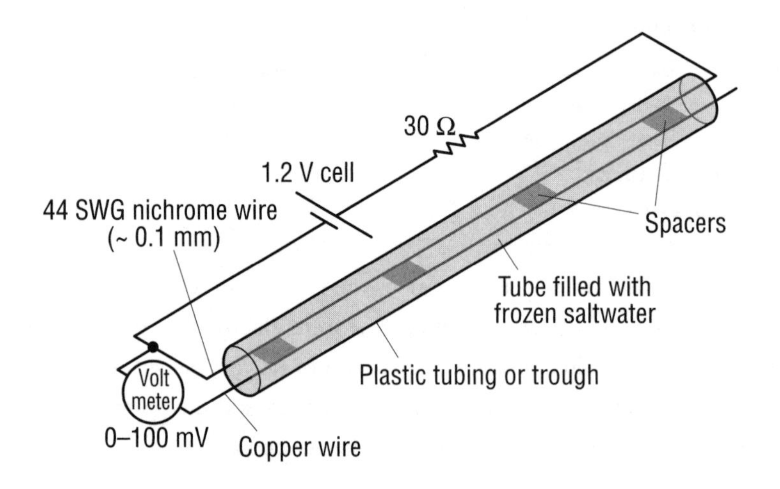

circuit diagram shows how the whole arrangement can be implemented. The small voltage on the output displays the position of the hot spot, while the current conducted indicates the severity of the hot spot.

The occurrence of phase transitions other than melting/freezing can also lead to changes in electrical conductivity. In particular, phase transitions in solids show increased conductivity, at least temporarily, as the transition occurs. This is often due to the generation of ions and electrons that are free to move during the rearrangement of the atoms from one crystal structure to another. Silver iodide (AgI) shows several such useful phase transitions (see Paul D. Garn, *Thermoanalytical Methods of Investigation*), and its electrical conductivity goes through a maximum at those phase transitions. AgI or a similar compound could perhaps be prepared as the solid "insulator" in coaxial cable and could then be used in much the same way as the freezer hot-spot detector just described.

Some solids also simply change their resistivity with temperature. Typical metal conductors increase gradually in resistance with temperature. This can be viewed as being due to increased scattering of the electrons that are trying to carry the current, thus requiring a higher propelling voltage, which manifests itself as increased resistance. Typical conducting oxides decrease in resistance with decreases in temperature, often steeply, as thermal excitation releases carrier ions and electrons. I once tried out a hot-spot detector for industrial burner flames based on the decrease in electrical resistivity of the insulator in an oxide-insulated heater element. (These are coaxial cables with magnesium oxide—based insulation and a nichrome central wire, all in an oxidation-resistant outer jacket.)

REFERENCES

Garn, Paul D. *Thermoanalytical Methods of Investigation.* New York: Academic Press, 1965.

Moore, Walter J. *Physical Chemistry.* 5th ed. London: Longman, 1972.

27 Electric Rainbow Jelly

> I procured me a triangular glass-prism, to try therewith the celebrated Phenomena of Colors. And in order therefore having darkened my chamber, and made a small hole in my window-shuts, to let in a convenient quantity of the sun's light, I placed my prism at its entrance, that it might be thereby refracted to the opposite wall. It was at first a very pleasing divertissement, to view the vivid and intense colors produced thereby.
>
> —Sir Isaac Newton, letter to Oldenburg, February 6, 1672

Human beings have always been fascinated by the relatively rare and pretty appearance of rainbows. Rainbows in the air are transient, which is perhaps where some of their magic lies. At the risk of losing part of that magic, you can make more lasting rainbows than Isaac Newton did. Here a specially made jelly turns into a solid rainbow when an electric field is applied for a minute or two. This is an easy trick to do simply, but it is quite difficult to arrange so that it looks as spectacular as Newton's "divertissement."

What You Need

- ❑ Gelatin
- ❑ Salt

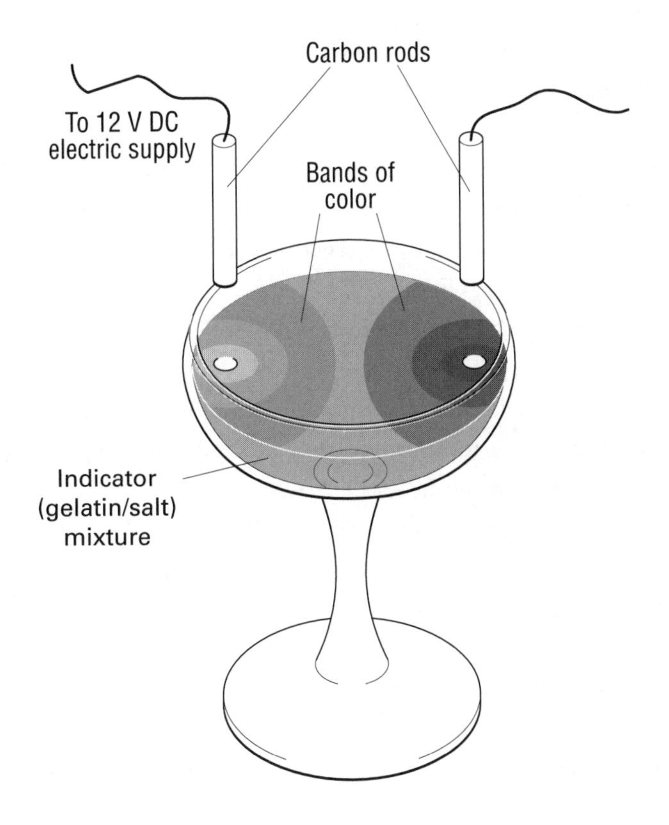

- ❑ Buffering agents (sodium bicarbonate? ammonium chloride + ammonia?)
- ❑ Universal indicator
- ❑ Glass or plastic containers
- ❑ Carbon rods (from a zinc chloride battery cell, for example)
- ❑ Piece of metal (I used a tin plate cut from an old cookie can.)
- ❑ 12 V or 6 V DC electric supply

What You Do

Prepare a strong gelatin solution with gelatin powder from your grocery store. Then dilute it with ten parts water and boil up the Universal Indicator (the BDH brand from Merck & Company works well, but anything a school science lab has will probably work). This is in an alcoholic solution, and adding water and heating it will dilute and evaporate some of the alcohol, which might otherwise stop the gelatin setting. Finally, add the two solutions together and pour into glass or transparent plastic containers. You should balance the amount of indicator added so that the color is not too dark a green. I think the effect is maximized in containers such as flat champagne glasses, which allow you to see the colors developed better. Let the gel set.

Now insert two pieces of carbon rod, one on either side of the gel. (You can extract the carbon rods from the middle of a cheap AA zinc-carbon battery: scrub them, and your hands, thoroughly afterward.) Now apply a DC voltage from a battery to the two electrodes in the gel. I used a 12 V car battery, but a 6 V battery would work just as well. Gas will be seen bubbling up at both ends at the higher voltages, forming a blob of colored foam around the electrode. There will also be a pungent smell, a strong "essence of swimming pool." Don't be tempted to apply the electrolysis current for two long: a couple of minutes or so is all that is needed. Now remove the electrodes.

Now wait . . . and wait . . . and wait. Over the next hours and days, a beautiful rainbow will develop, from red near the positive electrode (anode) to violet near the negative electrode (cathode). If you electrolyzed for too long you will soon have a gel with a red side and a violet side and a very thin band of darker colors in a line between them. If you have done it all right, you should have a rainbow of colors forming surfaces arranged concentrically around the spots where the electrodes were. Magically, purple, then blue, then green, perhaps a hint of yellow, extends from the metal electrode, while a red stain blending to orange extends from the carbon rod.

CAUTION

Mark the jelly clearly so that no one will try to eat it. "Electric Rainbow Jelly, do not eat" would do (although the foul salty taste will probably stop them from getting too far with eating it anyway).

THE SCIENCE AND THE MATH

The electricity is splitting up the salt in the gel into its constituent parts. Common salt is made of sodium and chlorine—sodium chloride. Salt dissolves by its constituent atoms splitting off into charged "ions"—atoms with an electric charge arising from an extra or a missing electron—which float around in the water. Positively charged sodium ions Na^+ float around an equal number of negatively charged chloride Cl^- ions.

The electric field of the electrodes pulls these apart, and they are attracted to the electrode with the opposite charge: Na^+ to the negative electrode (cathode), Cl^- to the positive electrode (anode). At the cathode, the Na^+ ions are discharged and form, together with the water in the solution, hydrogen gas and hydroxyl (OH^-) ions. At the anode the Cl^- ions are discharged and joined together in pairs to form chlorine gas (which gives the strong pungent smell) and hydrogen (H^+) ions. The chlorine gas, is, like every gas except oxygen and air, bad for humans, being poisonous, but it is produced in very small quantities here.

The electrolysis can be considered to inject electrons into the solution and neutralize hydrogen ions in the jelly at the metal electrode, releasing hydrogen and depleting the hydrogen ion concentration (thus giving an alkaline solution). The opposite process happens at the carbon (positive) electrode: electrons are removed, neutralizing hydroxyl (OH^-) ions and chloride (Cl^-) ions and releasing oxygen and chlorine respectively.

Positive (carbon) electrode: $4OH^- \rightleftharpoons O_2 + 2H_2O$

and $2Cl^- \rightleftharpoons Cl_2$ (becomes acid, that is, red)

Negative (metal) electrode: $2H^+ \rightleftharpoons H_2$ (becomes alkaline, that is, purple).

The chlorine released will "bleach"—destroy—a small amount of the indicator, leaving (if you look carefully) a small clear patch around the carbon electrode.

The story would end there but for the addition of the universal indicator and the gel. In a water solution, some of the water H_2O molecules split up into H^+ ions and OH^- ions. We say a water solution is acidic if it has a large number of hydrogen H^+ ions, and alkaline if it has a large number of hydroxyl OH^- ions. So sulfuric acid, which forms a large number of H^+ ions, is an acid, and potassium hydroxide, which forms a large number of OH^- ions, is an alkali.

Universal indicator is a mixture of chemical compounds that happen to have different colors, depending on the pH or, in other words, on how many hydrogen ions a solution has (how acidic it is).* One of the compounds in universal indicator changes from red at pH 4 to yellow at pH 6 via orange. Another compound changes from green to blue to violet. The result is that the indicator changes color in a spectrum from red to violet, just like the visible spectrum (although this is a result of the particular choice of indicator compounds, not a result of a fundamental law of nature). This color change happens when the pH of the solution turns from highly acid to highly alkaline. The indicator in the solution turns red (acidic) in the presence of hydrogen H^+ ions, and it turns violet (alkaline) in the presence of the hydroxyl OH^- ions.

The gel is a weak solid, but one through which the ions can move slowly by diffusion. The ordinary movements of the ions, due to their temperature, allow them to walk randomly but in general away from the electrodes where they formed. In this way, instead of a solid red color at the anode and solid violet at the cathode, the ions form a concentration gradient, with more ions at the electrodes but a decreasing concentration farther away. They thus form zones in which acidity is counteracted by alkalinity to varying degrees, and these zones are displayed in different colors by the indicator, because the indicator displays different colors depending on the acidity or alkalinity of the solution.

After the electrolysis, the hydrogen and other ions slowly diffuse through the jelly, spreading a pattern of colors across it, depending on the hydrogen ion concentration. The diffusion of the colors is slow, with the curious property that it goes more slowly as it goes farther afield. This follows from the diffusion equation whose law it follows. The diffusion equation for the concentration of compound C with respect to time t and distance (in one dimension) x can be stated as

$$\partial C/\partial t = D\partial^2 C/\partial x^2,$$

where D is the constant of diffusion. It follows from the fact that rate of diffusion is proportional to the concentration gradient $\partial C/\partial x$, and thus the rate of diffusion into a small elementary volume minus the rate of diffusion out is proportional to the gradient of that gradient, $\partial^2 C/\partial x^2$, and must be equal to the

*A reminder about pH, the unit of acidity: pH equals the negative logarithm of the number of hydrogen ions divided by the total number of water molecules in a solution. A strong acid like hydrochloric acid has a pH of about 1, pure water has a "neutral pH," a value of 7, while an alkali like sodium hydroxide has a pH of 14 or so.

rate of change of concentration in that elementary volume, $\partial C/\partial t$.

Solutions to the diffusion equation are many, but simple one-dimensional solutions are usually of form

$$C = kt^{-1/2} \exp(-x^2/4Dt),$$

where k is a constant. The equation shows a bell-shaped curve centered on the y-axis, whose width increases with the square root of time t, and with the height of the bell inversely proportional to the square root of time. (The behavior of diffusion can be considered on a microscopic scale to be the result of the $t^{1/2}$ dependence of the spreading out of individual molecules performing a "random walk." This analysis of a random walk is described in textbooks on kinetic theory like *Properties of Matter,* by B. H. Flowers and E. Mendoza, but it is easier and perhaps more instructive today to simulate on a desk-top computer.)

And Finally, for Advanced Users

You will no doubt be able to devise different shapes to put the jelly in before electrolysis: long thin columns, large flat plates. You can also choose different shapes of electrode: perhaps a central carbon rod and a circular surrounding metal electrode.

Try different shapes of electrodes made out of, for example, tin (from cans) or other metals. With these larger electrodes, where do the most intense colors form? Do the edges of the color bands follow lines of equal voltage (equipotentials)?

The colors may be changed, to some extent, by dissolving metal ions. You could use a carbon rod for one electrode if this is troublesome. Iron at the cathode, for example, tends to give a yellow color from the Fe^{3+} ions there, formed as the iron atoms are ionized.

The use of buffer solutions gives a different background color to the jelly. The addition of small amounts of buffers also has a beneficial influence on the artistic effect because they increase the distance over which the intermediate colors are seen. A typical buffer solution is a salt of a strong acid and a weak alkali with excess alkali, or a weak acid with a strong alkali and excess acid. Examples are respectively ammonium chloride and ammonia mixture, or sodium citrate with citric acid. Buffer solutions change only slightly in pH, even with the addition of relatively large amounts of acid or alkali.

REFERENCES

Flowers, B. H., and E. Mendoza. *Properties of Matter.* Chichester, UK: Wiley and Sons, 1970.

Hakfoot, Casper. "Newton's Optics: The Changing Spectrum of Science." In John Fauvel, Raymond Flood, Michael Shortland, and Robin Wilson, eds., *Let Newton Be!* Oxford: Oxford University Press, 1988.

Infernal Inventions

28 *Binary Match*

Double, double toil and trouble,
Fire burn and cauldron bubble
Eye of newt, and toe of frog
Wool of bat, and tongue of dog
Adder's fork, and blind-worm's sting,
Lizards' leg, and owlet's wing,
For a charm of powerful trouble,
Like a hell broth boil and bubble.

—William Shakespeare, *Macbeth*

Before the days of cigarette lighters or matches, lighting a fire was a problem. Although the bow-and-arrow system and the iron-and-flint system were available, they were not easy, especially if the weather was damp. As soon as powerful chemicals became available, toward the end of the eighteenth century, various systems were devised that did not involve any mechanical effort or ingenuity on the part of the user. John Emsley describes some of them in *The Shocking History of Phosphorus*. Simply by mixing a powerful sulfuric or nitric acid with a suitable combustible material, the user could reliably obtain a powerful flame. The acids were sealed in small glass vials, with the combustible wrapped in waxed material to keep it dry. However, it must have required some nerve to carry a number of these on your person. Personally, I think I would

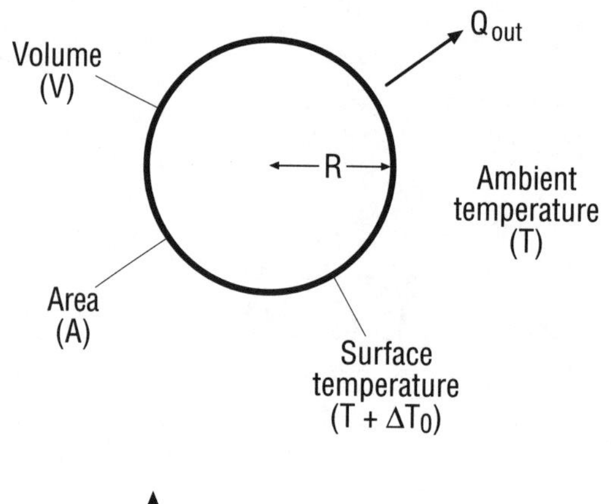

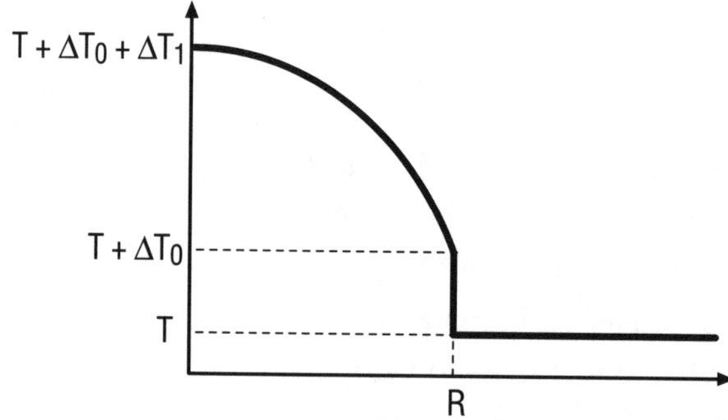

prefer Damocles' sword suspended over my head to a set of sulfuric vials in my pants pocket!

Our trick here is a way of achieving Binary Match function using only two fairly innocuous ingredients. One is syrupy glycerol (often called glycerin, sold for use in cooking), and the other is the readily available oxidizing agent potassium permanganate. Glycerol is technically a trihydric alcohol, which has three -OH side groups. It is the backbone of animal and vegetable fats, which comprise fatty organic acids like stearic or oleic acids, joined to each of the -OH side groups of the glycerol. It is obtained as a by-product of making soap.

What You Need

- ❑ Potassium permanganate
- ❑ Glycerin
- ❑ Concrete
- ❑ Safety goggles

What You Do

Make a small conical heap of the purple potassium permanganate crystals about 1.5 cm across. (If the crystals are bigger than 1 mm or so, they will need to be crushed to a smaller average size, for instance, with the back of a spoon on a saucer.) Make a small crater in the top of the heap and, using a teaspoon, add half a teaspoon of glycerol.

If all is well, the little heap of crystals will begin to bubble and a little wisp of vapor will begin to rise. The mixture will then start to bubble furiously. The bubbling will continue for a minute or so, and then a wisp of smoke will rise, until finally a white flame will burst out and consume much of the heap in a few seconds. (Perhaps Shakespeare's witches in *Macbeth* should have tried permanganate and glycerol rather than the eye of newt, toe of frog, and so on.)

WARNING

The smoke given off is not too unpleasant, but this trick is nevertheless best performed outside. You might set fire to something, so make sure that there is nothing flammable nearby, and that there is a large piece of something heat resistant (like a large slab of concrete) underneath. Wear goggles in case the mixture spits out a hot piece of permanganate. Don't use more than a teaspoon of permanganate.

THE SCIENCE AND THE MATH

The heat generated by the reaction of the potassium permanganate with the glycerol is sufficient to raise the temperature. This increases the reaction rate, which in turn raises the temperature, which increases the reaction rate, and so on. After a few seconds, the system has achieved "thermal runaway," in which the temperature shoots up quickly to the ignition point of glycerol.

Potassium permanganate oxidizes the $-CH_2OH$ group of most alcohol molecules to an organic acid -COOH group. Glycerol works especially well because it has three -OH groups in each molecule, giving a high heat of reaction per unit weight with the permanganate.

If the amount of the two reagents is too small, then ignition will not occur. There is something akin to a critical mass, a concept, as mentioned before, first applied to the mass of uranium required for a nuclear explosion. The concept is complicated here by the effect of the size of the permanganate crystals, which affects the rate of heat evolution.

The critical-mass concept is most simply approached by noting that heat (or neutron) loss Q_{out} is proportional to surface area A, while heat (or neutron) generation Q_{in} is proportional to volume V.

$$Q_{in} = Q_v V = Q_v (4\pi/3) R^3,$$

where Q_v is the heat generated per second per unit volume. Q_v is not a constant—it varies with temperature, as we will see—but let's assume for the moment it is constant for small rises in temperature.

The heat loss is also proportional to the temperature above the surroundings ΔT_0, a fact sometimes grandly entitled Newton's law of cooling, giving

$$Q_{out} = cA\Delta T_0 = c4\pi R^2 \Delta T_0,$$

where c is a constant.

So, equating the two, assuming equilibrium,

$$\Delta T_0 = Q_v R/(3c),$$

where R is the radius of the heap (assumed spherical for these calculations). There is a further complication, however: the temperature in the middle of the heap of chemicals will be much hotter than the outside—its thermal conductivity is relatively low. In fact, the temperature ΔT_i above the surface inside a long, cylindrical chemical reactor peaks in the middle, following a square law:*

$$\Delta T_i \sim Q_v(R^2 - r^2)/6K,$$

where r is the radius at which the temperature is measured, and K the thermal conductivity.

So the equation

$$\Delta T_{tot} = \Delta T_0 + \Delta T_i = (Q_v/3c)R + (Q_v/6K)R^2$$

gives the total temperature rise ΔT_{tot} from the hottest place (in the middle) relative to the ambient air around. Of course, Q_v is not constant but varies quite rapidly with temperature, following a Boltzmann factor:

$$Q_v = C \exp(-E/kT),$$

where C is a constant, E an energy, which can be thought of as the "activation energy" (the energy required for one molecule to overcome a repulsive force barrier and approach another molecule close enough to undergo reaction), k Boltzmann's constant, and T the absolute temperature. Unfortunately for the analysis given, with typical values of E at room temperature, this Boltzmann equation gives a factor of 2 for each 10°C temperature rise. The reader is therefore invited to carry out some further analysis here: to factor the temperature variation of Q_v into the equations just provided to give a more accurate analysis of where a high temperature rise is seen.†

*This kind of effect is found in large haystacks and in large heaps of coal. In the presence of a little moisture, the hay or coal may begin to oxidize, and although this may involve a surface temperature rise of only a few degrees, the center of the stack or heap may be at hundreds of degrees. This may cause a serious fire if nothing happens to cool the center. On a smaller scale, linseed oil, used in paints and varnishes, oxidizes slowly in the air exothermically, and rags soaked in it can heat up. Fires have occurred from this cause from time to time, including, for example, the infamous 1991 fire of the skyscraper One Meridian Plaza in Philadelphia.

†I once ran tests on a chemical reactor, a cylinder filled with a ceramic-based catalyst, which, unfortunately, obeyed this equation, or something like it. The outside of the reactor got hot, but not hot enough to cause alarm. But the system mysteriously started to fail. After much head scratching, and the observation that a small plastic pipe had been overheated, we cut open the reactor and were astounded at what we found. The middle was completely fused: it must have gotten white-hot, it had overheated so much.

And Finally, for Advanced Users: Critical Mass

Why does it make any difference if you make the permanganate crystals smaller? The Binary Match would be much more convenient as a practical source of fire if it could be made smaller. But does the Binary Match ignite in smaller quantities? Or is there really a critical mass? Given thermal capacities, reaction rates at a particular permanganate particle size, and heats of combustion, can you calculate the critical mass for a sphere of evenly mixed material?

The concept of a critical mass arose when it was realized that uranium-235 metal would simply explode if you jammed enough of it together in a tight space for a few milliseconds. Why is this? A single ^{235}U atom will occasionally explode (called "fission," technically) and emit a neutron. These neutrons are quite likely to strike and cause two other ^{235}U atoms to explode and emit their own two neutrons. There are now four neutrons, which go on to cause the fission of four further ^{235}U atoms. A chain reaction has occurred, one that—if all the emitted neutrons strike and cause fission—will cause an exponential increase in exploding atoms and, within a few milliseconds, a bulk explosion of all of the uranium and a nuclear explosion.

However, this chain reaction will not occur if many of the neutrons escape before striking ^{235}U atoms to cause fission. With a small ball of ^{235}U, that is what happens: the neutrons, which travel a few centimeters before hitting an atom, tend to escape from the surface before hitting an atom, on the average. The critical mass is the mass of the ^{235}U ball that is just big enough to retain enough neutrons to cause an exponential chain reaction—it was first calculated in 1940 to be about 8 kg. There are similar effects in the Binary Match: for uranium, read permanganate/glycerol; for neutrons, read heat.

REFERENCES

Emsley, John. *The Shocking History of Phosphorus.* London: Macmillan, 2000.

Faith, Nicholas. *Blaze, the Forensics of Fire.* London: Channel 4 Books / Macmillan, 1999.

29 *Ultimate Bunsen Burner*

With a great noise, the elements shall melt with fervent heat.

—2 Pet. 3:10–12

Obviously, Saint Peter, quoted in the epigraph, had seen early versions of the Ultimate Bunsen Burner described here. It does indeed create a great deal of noise, and it certainly melts most elements.

After tantalizing short pieces have appeared on the subject in the *New Scientist* and *Die Welt* and on the BBC, readers all over the world must be panting for the details of how to build their own Ultimate Bunsen, a project I have worked on, intermittently, for years. With the aid of a subtle design incorporating a pre-heater assembly, this burner raises the performance of a simple propane/air burner to oxyacetylene levels, allowing the user to melt glass, quartz, and even steel plate.

A project like this must inevitably be buffered with strong safety warnings, and in any case the project requires access to lathe and metal-welding facilities. However, a college class with workshop facilities may find this a fascinating glimpse into the future of burners.

Although few readers will have the time or the equipment to do this project, it is a stimulus to thought, and I think it is worth including even if it remains a thought experiment for most readers. It shows that even in subject areas as mature as burners, there are still new developments. And if *Scientific American*

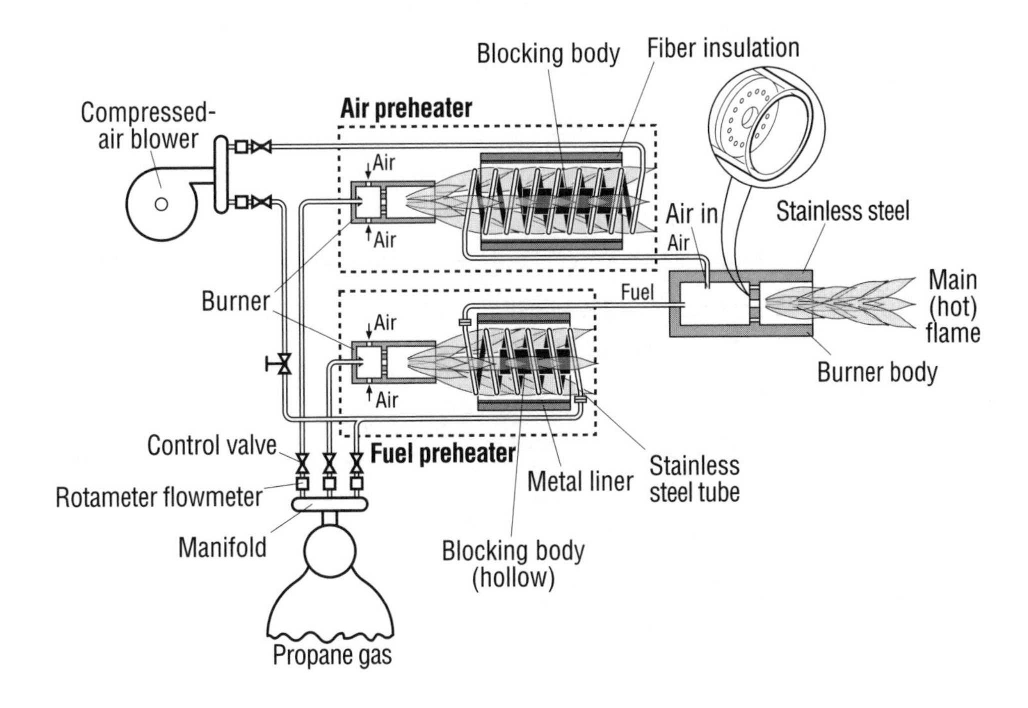

can include in its "Amateur Scientist" column (a 1972 issue, I believe) a build-it-yourself "death-ray" carbon dioxide laser capable of cutting holes in things à la *Star Trek*—a project of truly prodigious difficulty and with even more serious safety implications—then I can include a gas burner, even one that can burn through steel!

Here we supply a burner with air and fuel, but we preheat both the air and fuel to high temperature, so that the flame starts from a 600°C or so advantage relative to a standard flame. With such a flying start, and particularly because the fuel gas partly "cracks" to form lower-molecular-weight gases like hydrogen—which burn more quickly and vigorously—this burner will get hotter than any other gas/air burner and will easily melt its way through steel plate up to 3 mm or so thick—a whole lot thicker than anything in an automobile!

The Degree of Difficulty: Extremely Difficult and Dangerous

This is a very difficult project, to be attempted only by qualified persons at universities and colleges. (See the caution on safety that appears later.)

What You Need

- ❑ Burner head (machined out of 316 L stainless steel)
- ❑ 2 commercial propane burners, about 20–25 mm diameter (for preheat)
- ❑ 2 preheat coils
- ❑ Jets (Between 0.4 and 0.8 mm are useful sizes.)
- ❑ 2 flow-control valves
- ❑ 1 fine-control valve
- ❑ Swagelok fittings
- ❑ 1/4″ 316 L stainless-steel tubing
- ❑ Propane fuel gas in a cylinder
- ❑ Propane gas regulator
- ❑ Air compressor and regulator or high-pressure (100 mbar centrifugal) fan
- ❑ High-quality rubber gas hose
- ❑ Match or other source of ignition
- ❑ Ceramic fiber insulation (optional)

What You Do

Machine the burner head as per the drawing. Bend the two preheat coils and attach them by welding (the gas connection is best welded so that it will be leak tight even after severe temperature cycling). Add a suitable jet, and add the insulating fiber if desired (this makes the burner more efficient). Arrange the preheater burner to heat the coils. Then open the gas valve and light up the main burner to give a large (300–500 mm long) flame. Open the air-bleed valve to give 1 liter/min airflow.

Now turn on the preheater flame and allow it to warm up for five minutes. Then turn up the main burner flow (both gas and air) to the maximum sustainable before flame liftoff (keep a small pilot flame burning below the burner while you do this).

You should now find that you have an intense 30 mm flame that will instantly heat wire to white heat, and that will melt through steel plate up to about 3 mm thick. Watch out for the stream of white sparks (actually tiny droplets of burning steel) that the burner will blow out of the steel.

The Tricky Parts

The burner must warm up before high-flow operation is attempted. The flow valves should all be adjusted to relatively low flow rates until the preheater has done its work.

Any experimental gas burner is dangerous. A burner capable of melting metals like steel has clear dangers. Although it is much safer than an oxy-acetylene burner, employing as it does only propane and air, like any gas appliance it is a potential source of hazard. It should only be used in a well-ventilated place, because, depending on how it is used, appreciable amounts of products of incomplete combustion and nitrogen oxide (NO_x) gases will escape from the flame into the atmosphere. Keep a fire extinguisher handy, and place the gas shut-off valve in a handy place too. The whole burner gets very hot—this should be obvious, as large parts of it glow bright orange in operation! However, like those rings on top of an electric stove, the metal parts do remain hot enough to give a small burn wound on your skin for five or ten minutes after operation. Finally, if unburned gas builds up in a small room and is then ignited, there may be an exceedingly dangerous gas/air deflagration or even detonation explosion.

CAUTION

Use extreme caution in carrying out this experiment. There is a danger of fire, burns, and gas explosions. This project should be attempted only by qualified adults.

The Surprising Parts

Most people would say that heating the fuel gas is a waste of time: it is only a few percent of the airflow, and so heating it would appear to be a waste relative to simply heating the air a little hotter. The first surprise is that it is not a waste of time, and heating the fuel does make a difference. The reason for this lies in the fact that the fuel is being cracked to higher energy species (including, for example, free radicals and acetylene) as well as being heated, and that the cracked species have a higher flame speed.

The second surprise is that heating the fuel does not immediately coke the burner with a thick filling of soot. The air bleed prevents this, by causing cracking to occur to oxygenated hydrocarbons or CO rather than carbon (soot).

The third surprise is that the flame shrinks in size so much. This is despite a factor of three times or more increase in flow rate (the air and gas expand approximately proportional to absolute temperature by this much). The reason for this lies with the increasing flame speed at higher temperatures and with the partially cracked fuel gas, which contains gases with very high flame speeds such as acetylene and hydrogen.

The fourth surprise is that the burner does not usually melt itself down. Curiously, the heated gas and air supplied, being cooler than the flame and supplied under pressure, actually act as a coolant to prevent the flame radiant heat and convective flow from melting the burner parts. The nozzle tip is the only part prone to melting, and even this tends to survive because the oxide formed on its surface has a high chromium content and so acts as a heat barrier.

Using the Ultimate Bunsen

The flame created here will be powerful enough to melt ordinary grades of steel (up to the 1540°C, the melting point of pure iron) and even platinum (1,770°C), along with all the more easily melted metals, such as lead, brass, copper, nickel, silver, and so forth.

Material in rod or wire form is most easily melted, but sheets and small blocks can be melted too, with patience. In the case of small parts, it is difficult to hold them—steel tongs, for example, will themselves melt. The use of ceramic such as aluminum oxide—alumina—helps here.

The torch is very useful for melting and bending glass tubes and other items made in the high-temperature glasses such as Pyrex. *Beware:* Small amounts of metal scale may break off the inside of the burner parts during start-up, and these little particles can end up sticking to the glass surface. (This is the principle used for its beneficial effects in industry, marketed as "flame spraying.")

THE SCIENCE AND THE MATH

The burner combines a number of different effects to achieve its formidable heat.

First, it heats the combustion air by 500°C, raising the flame temperature by an amount that is a sizeable fraction of that 500°C. This raises the temperature of the flame, although not by as much as 500°C. The apparent loss of heat arises because the hotter final flame has a greater degree of disassociated molecules—free radicals, free atoms—than a standard cooler flame. The disassociation of the molecules takes away some of the energy put in by the preheat.

Second, it heats the fuel to a high enough temperature that it cracks into a mixture of gases such as hydrogen and ethylene—gases whose flame speed is much higher than that of propane itself. Propane reacts relatively slowly with air: flame speeds of around 40 cm/s are typical, while hydrogen/air flame speeds peak at over 250 cm/s (see, for example, G. Monniot, *Principles of Turbulent Fired Heat*).

The higher the flame speed, the more swiftly the gas/air mixture will burn, combusting in a smaller volume and giving smaller losses from conduction and convection: gases with higher flame speeds give rise to a more intense flame. The flame is shortened greatly relative to a normal flame with the same gas flow and area. The high intensity also means that the flame is not cooled by the work piece.

Third, once the gases get to the burner nozzle, more effects come into play. The higher flame speed allows the use of very high flow rates, which create very high turbulence, which allows even higher flow rates, and finally, this high turbulence effects better heat transfer to the work piece. (Readers are referred for further details to the published patent or to the author via the publisher.)

The Electric Burner

Various associates and I have more recently been working on electric preheat versions of the torch, in which simple nichrome spirals are used to heat the gas. The latest demonstration version of the torch uses three small (3 cm diameter × 10 cm) heater units, all identical ceramic formers wound with a NiCr wire spiral (actually a helical "spring" wound in a spring shape itself). The heaters are enclosed in a transparent fused-quartz tubing assembly, which feeds the main burner, made, as usual, of stainless steel. The heaters are fed by their own 5 A-maximum TRIAC regulators. Although only low air and gas pressures (a few tens of millibars) can be used in this torch, it still works well. It has less well-developed turbulence than earlier torches but is also much quieter in operation.

Careful design is needed to ensure that gas flow is very evenly distributed through the spiral, and that no cool-gas "channeling" occurs through the preheater assemblies. Channeling also causes hot spots, and hot spots lead to overheating and burnouts.

And Finally, for Advanced Users

The use of further rotameter flow measurement meters on the main burner gas and air lines is helpful for tuning the burner to optimum conditions and should help you to reach the same conditions each time the burner is used.

The lifetime of the burner as given is limited by the stainless-steel nozzle parts (which are easily replaceable) and to a lesser extent by the preheat coils. With replacement of nozzle-tip parts, the burner should run for hundreds of hours without trouble. The preheat coils can also be replaced (the gas preheater is less easy on account of the necessity to weld the end of it). I am working on increasing the running life by a more subtle choice of materials. For example, we are looking into the increased coil lifetime given by nichrome and Inconel alloys.

Another possibility is the addition of a further heating coil arranged to give hot air from a small (about 1 mm or less) orifice near the burner flame tip. This will allow you to carry out crude steel cutting—at a rate of up to 100 mm per minute or so. The hot air blast is needed because cold air might blow away molten steel from the cut line but would also cool the melt and prevent it from being blown away.

This is a device that I and various associates are still working on, and there is some prospect of commercial burners based on these principles becoming available in a couple of years.

REFERENCES

Downie, N. A. "A New Bunsen Burner." *New Scientist*, April 1998.

———. "Improvements in Burners." World Patent (PCT) #WO09901698A1, 1999.

Monniot, G. *Principles of Turbulent Fired Heat.* Paris: Gulf Publishing, 1985.

Useful Materials and Components

Most households and school or college workshops have abundant facilities for all the projects described in this book. However, here are a few tips on what I have found helpful in the way of materials and components. There are some tools and parts that are less obviously useful that I have found to be particularly efficacious.

INSTRUMENTS AND TOOLS

Electrical multimeter. From about $10 you can get what is really quite a sophisticated digital display multimeter. The price is so low that it is not worth trying to manage without one. For slightly more expense, you can get a multimeter with useful additional functions like thermocouple readout or frequency measurement.

Oscilloscope. Even a simple, crude oscilloscope is helpful in debugging electronic circuits. This used to be a very expensive item. However, today you can buy, for $20 or $30, an oscilloscope simulator, a small box you plug into your home computer that allows the computer to offer oscilloscope (and chart recorder) functionality.

Electronic kitchen scales. These scales are helpful where weights or forces must be measured with reasonable accuracy.

Stroboscope. For stroboscopic viewing, a small disco stroboscope (from about $15) or a disk with slots on a geared motor mounted on a stick (viewing objects through the slots will give as good a result as the stroboscope in many cases). A wheel-brace-style hand drill, especially the pistol-grip type, is also a useful way of mounting the disk.

Glue. I recommend a hot-melt glue gun (sets in thirty seconds or so, but don't burn your skin with it!) or cyanoacrylate (sets in two or three minutes, but don't glue your skin with it!) for gluing. Gluing is quicker and easier than many alternative methods of joining, and traditional wood glues and two-pack epoxy adhesives are as good as ever. However, I have found these two kinds of rapid-setting glues particularly helpful.

MATERIALS

Rosin. Powdered rosin (as sold for ballet dancing) is very useful for increasing the coefficient of friction between surfaces.

Tape. As well as the usual cellophane (Scotch) tape, cloth-reinforced duct tape (aka "duck tape") is very useful. A small can or bottle of a solvent, for degreasing surfaces, is a useful adjunct to the tape, as the tape adhesion is greatly improved when the surface is thoroughly cleaned. I often use the solvent sold for use with cellulose car paint.

COMPUTERS

Computer software. A financial spreadsheet program is useful not only for monetary calculation. I have found these very effective for scientific simulations. You can easily set up numerical calculations of many phenomena, without any real computer skills other than knowledge of the spreadsheet syntax. For example, set a series of time intervals down the left-hand column, enter the elements of a differential equation as finite difference formulas in subsequent columns, and you can plot or solve differential equations.

PARTS

Large electronic-component distributors are obviously good people to buy electronic parts from. Less obviously, they are also, in many cases, convenient suppliers of many nonelectronic goods, from electric motors to tools.

Electric motors. I favor small DC electric motors that are made, or used to be made, for the operation of portable cassette tape recorders. The motors I have in mind have a round metal body 28 mm long and 23.8 mm in diameter, work on voltages between about 0.6 and 5 volts, and have a 2 mm diameter shaft. They have the advantage of being about the smallest motor that shows a reasonable degree of electrical efficiency (50 percent) and draw (compared to other designs) a reasonably low current from a battery supply, around a quarter of an amp at 1.5 V. An equally common motor design has a slightly smaller body, 25 mm long, which has flats on the sides, draws several times the current (1 A at 1.5 V) and is less efficient (30 percent), although it does give more power and is more common in small toys and models.

Wheels. It is convenient to have wheels that simply jam onto the end of the motor. For a 2 mm shaft, that means a 1.9 mm diameter hole. Failing that, you can make a 2 mm hole in the wheel, then abrade the motor shaft, and then glue on the wheel using cyanoacrylate glue. You can jam a wheel that has too large a hole in the middle onto the motor shaft by using a short piece of tubing, such as insulation taken from a suitable piece of electric cable. You may find that the flexible insulation on 25 A household copper cable is about right for doing this.

Transmissions. Electric motors run too fast for many potential applications, and a reduction transmission is needed. Short of making your own, which it is hard to do correctly (although see the Flying Pulleys project), it is difficult to know what to recommend. It is expensive to buy a model transmission, while transmission units made from Meccano or Erector Set parts are bulky and heavy—and expensive, if you don't recycle the parts. Some schools' suppliers stock transmissions or components for making them, such as simple worm drives, but only a few keep units that are truly effective and low cost. Electric screwdrivers can be bought at a very low price and have a splendid, well-built gearbox and motor unit inside them. Finally, it is always worth examining discarded toys to see if the transmission or electric motor unit will be useful.

Meccano and Erector Sets. Although it is possible to build useful demonstrations using only parts from the set, I find that it is invariably more efficient to use a few parts from these for one or two vital functions and to rely on inexpensive materials like wood for the bulk of the device.

Shelf brackets. Small, pressed-steel shelf brackets are a boon for constructing projects quickly. By attaching an upright post to a wooden base, for example, you have an instant but very inexpensive equivalent of the conventional chemistry lab clamp stand.

Electric-motor speed control. I think that the easiest way to control the voltage to an electric motor is to vary the number of battery cells connected in series to supply it. In this way, the voltage supplied can be adjusted in 1.2 V (rechargeable NiCd cells), 1.5 V (alkaline cells), or 2 V increments (lead acid cells). Using the same philosophy, a silicon diode (a power diode with an appropriately generous current rating, such as a 2 A diode for a 1 A motor) can be used in series with a number of battery cells and has the effect of reducing the voltage by about 0.6 V. In this way, supply voltage can be adjusted in 0.6 V increments.

A simple variable-resistor control, as used for model slot cars, is another possibility, but one that will work only for motors of the correct voltage and current consumption (around 12 V and less than 1 A or so). An even simpler alternative with a high power capability is just a length of nichrome resistance wire, which can be loosely coiled up to make it neater.

Probably the best solution, offering motor control for motors with a wide range of voltage and current demand, is a pulse-width control system. This is simply a transistor that is switched on and off a few thousand times per second, with the proportion of on-time determining the speed. A simple circuit for pulse-width control is available in manuals of electronic circuits; see, for example, Rudolf F. Graf and William Sheets, *Encyclopedia of Electronic Circuits* (6:415, fig. 57-8).

Batteries and electric power supplies. I find that the most satisfactory battery cells are NiCd cells, with the AA size being the cheapest and often perfectly satisfactory. These can be recharged hundreds of times (although not forever; they do tend to die after a few years) and offer high peak-current capability. Beware that battery holders usually use steel spring connectors, which can get very hot in applications that use the peak-current capability of the NiCd cells. In that case, you should solder wires to the top of the steel spring so that the spring is no longer in the circuit, or devise a way of jamming the connecting wires directly against the cell terminals.

Power supply units working off the main house current are often unsatisfactory. Many have insufficient power capability, being designed to supply electronic circuits rather than electric motors. Others have too high a voltage to be useful: many small electric motors need less than 2 volts in typical applications. Exceptions are transformer units designed for use with 12 V model trains and slot cars, and the switch-mode power supplies that can be salvaged from old desktop computers, which typically have a powerful 5 or 5.6 V output.

A Reminder about Units

Throughout this book I have used mainly metric measurement units. These are now the standard units internationally and are certainly more useful for calculation. But old-fashioned English units often give some people (myself included) a better feel for magnitudes of quantities, so here are some reminders about units.

25.4 mm	=	1″
300 mm	~	1 ft
1 m	~	3 ft 4″
1.609 km	=	1 mile
16.4 cm^3	=	1 cu. in
28.3 liters	=	1 cu. ft
28.35 g	=	1 oz
0.454 kg	=	1 lb
0.3048 m/s	=	1 ft/sec
1.609 km/h	=	1 mph
1.055 kJ	=	1 BTU
0.745 kW	=	1 horsepower
Temp (C)	=	5/9 (Temp (F) – 32)

Bibliography

Alexander, R. McNeill. *Exploring Biomechanics.* New York: W. H. Freeman and Scientific American, 1992.

Atkinson, Norman. *Sir Joseph Whitworth.* Stroud, UK: Sutton, 1996.

Bader, P., and A. Hart-Davis. *Local Heroes: The Book of British Ingenuity.* Stroud, UK: Sutton, 1997.

Baker, R. *New and Improved.* London: British Library, 1976.

Banks, Robert. *Towing Icebergs, Falling Dominoes.* Princeton, N.J.: Princeton University Press, 1998.

Braddick, H.J.J. *Vibrations and Waves.* New York: McGraw-Hill, 1965.

Bunch, B., and A. Hellemans. *The Timetables of Science.* London: Sidgwick and Jackson, 1989.

Burke, James. *Connections.* London: Macmillan, 1978. ("Connections" is also the title of Burke's monthly column in *Scientific American.*)

Dale, Rodney, and Joan Gray. *Edwardian Inventions.* London: W. H. Allen, 1979.

DeBono, Edward. *Lateral Thinking.* New York: Harper and Row, 1970.

Denny, Mark W. *Air and Water.* Princeton: Princeton University Press, 1993.

de Vries, Leonard. *Victorian Inventions.* London: John Murray, 1971.

Dewdney, A. K. *The Turing Omnibus.* New York: Computer Science Press, 1989.

Doherty, Paul, and Don Rathjen. *The Exploratorium Science Snackbook Series.* New York: Wiley, 1991–1996.

Douglas, J. F., J. M. Gasiorek, and J. A. Swaffield. *Fluid Mechanics.* London: Pitman, 1979.

Downie, Neil A. *Industrial Gases.* London: Chapman and Hall, 1997.

Dummer, G.W.A. *Electronic Inventions and Discoveries.* Oxford: Pergamon Press, 1978.

Ehrlich, Robert. *Why Toast Lands Jelly-Side Down.* Princeton: Princeton University Press, 1997.

Elmore, William C., and Mark A. Heald. *Physics of Waves.* New York: McGraw-Hill, 1969.

Fleming, J. A. *Electric Wave Telegraphy and Telephony.* London: Longmans, Green, 1916.

Gardner, Martin. *Science Tricks.* New York: Sterling, 1998.

Graf, Rudolf F., and William Sheets. *Encyclopedia of Electronic Circuits.* Vol. 6. New York: McGraw-Hill, 1996.

Haldane, G.B.S. *On Being the Right Size.* Oxford: Oxford University Press, 1985.

Hankins, Thomas L., and Robert J. Silverman. *Instruments and the Imagination.* Princeton: Princeton University Press, 1995.

Herbert, T. E., and W. S. Procter. *Telephony.* 2 vols. London: Pitman, 1934.

Hill, Winfield, and Paul Horowitz. *The Art of Electronics.* Cambridge: Cambridge University Press, 1989.

Jewkes, John, David Sawers, and Richard Stillerman. *The Sources of Invention.* London: Macmillan, 1958.

John, Fleming, and John Lenihan. *Science in Action.* Bristol, UK: Institute of Physics Publishing, 1979.

Jones, David E. H. *The Inventions of Daedalus: A Compendium of Plausible Schemes.* Oxford: W. H. Freeman, 1982.

Jones, D.R.H. *Engineering Materials 3: Materials Failure Analysis.* Oxford: Pergamon Press, 1993.

Kaye, G.W.C., and T. H. Laby. *Tables of Physical and Chemical Constants.* 16th ed. Harlow, UK: Longman, 1995.

Kamm, Lawrence J. *Designing Cost-Efficient Mechanisms.* New York: McGraw-Hill, 1990.

Kibble, T.W.B. *Classical Mechanics.* 2d ed. Maidenhead, UK: McGraw-Hill, 1973.

Kirk-Othmer Encyclopedia of Chemical Technology. 36 vols. 3d ed. New York: Wiley Interscience, 1985. (An incomplete 4th ed., 1999, is also available.)

Laithwaite, Eric. *An Inventor in the Garden of Eden.* Cambridge, UK: Cambridge University Press, 1994.

Maunder, Leonard. *Machines in Motion.* Cambridge, UK: Cambridge University Press, 1986.

Messadie, Gerald. *Great Modern Inventions.* Edinburgh: W. and R. Chambers, 1991. (This book contains some spectacular mistakes, the results of a hasty translation from the original French.)

Michel, Walter J., J. Peter Sadler, and Charles E. Wilson. *Kinematics and Dynamics of Machinery.* New York: HarperCollins, 1983.

Morus, Iwan Rhys. *Frankenstein's Children: Electricity, Exhibition, and Experiment in Early Nineteenth-Century London.* Princeton: Princeton University Press, 1998.

Ord-Hume, Arthur W.J.G. *Perpetual Motion.* London: George Allen Unwin, 1977.

Richardson, E. G. *Physical Science in Art and Industry.* London: English Universities Press, 1940.

Riley, K. F. *Problems for Physics Students.* Cambridge: Cambridge University Press, 1982.

Sharpe, Carill, ed. *Kempe's Engineers' Yearbook.* Tonbridge, UK: Miller-Freeman, 1996.

Silverman, Mark P. *Wave and Grains.* Princeton: Princeton University Press, 1998.

Sobel, Dava. *Longitude.* London: Fourth Estate, 1996.

Stewart, Ian. *Game, Set, and Math.* Oxford: Basil Blackwell, 1989.

Thomson, N., ed. *Thinking Like a Physicist.* Bristol, UK: Adam Hilger, 1987.

Thorpe, Nick, and Peter James. *Ancient Inventions.* London: Michael O'Mara Books, 1994.

Tyndall, John. *Fragments of Science.* 8th ed. London: Longmans, 1892.

Vogel, Steven. *Life in Moving Fluids.* 2d ed. Princeton: Princeton University Press, 1994.

Walker, Jearl. *The Flying Circus of Physics.* New York: Wiley, 1975.

Wightman, W.P.D. *The Growth of Scientific Ideas.* Edinburgh, UK: Oliver and Boyd, 1950.

Index